Frontiers in Mathematics

Michael I. Gil'

Stability of Vector Differential Delay Equations

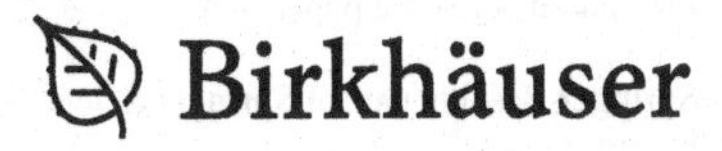

Michael I. Gil'
Department of Mathematics
Ben Gurion University of the Negev
Beer Sheva, Israel

ISSN 1660-8046 ISSN 1660-8054 (electronic)
ISBN 978-3-0348-0576-6 ISBN 978-3-0348-0577-3 (eBook)
DOI 10.1007/978-3-0348-0577-3
Springer Basel Heidelberg New York Dordrecht London

Library of Congress Control Number: 2013932131

Mathematics Subject Classification (2010): 34K20, 34K06, 93C23, 93C05, 93C10, 34K13, 34K27

Cover design: deblik, Berlin

Printed on acid-free paper

Springer Basel is part of Springer Science+Business Media (www.birkhauser-science.com)

Contents

Preface

1. The core of this book is an investigation of linear and nonlinear vector differential delay equations, extended to coverage of the topic of causal mappings. Explicit conditions for exponential, absolute and input-to-state stabilities are suggested. Moreover, solution estimates for these classes of equations are established. They provide the bounds for regions of attraction of steady states. We are also interested in the existence of periodic solutions. In addition, the Hill method for ordinary differential equations with periodic coefficients is developed for these equations.

The main methodology presented in the book is based on a combined usage of recent norm estimates for matrix-valued functions with the following methods and results:

a) the generalized Bohl–Perron principle and the integral version of the generalized Bohl–Perron principle;
b) the freezing method;
c) the positivity of fundamental solutions.

A significant part of the book is devoted to a solution of the Aizerman–Myshkis problem and integrally small perturbations of linear equations.

2. Functional differential equations naturally arise in various applications, such as control systems, viscoelasticity, mechanics, nuclear reactors, distributed networks, heat flow, neural networks, combustion, interaction of species, microbiology, learning models, epidemiology, physiology, and many others. The theory of functional differential equations has been developed in the works of V. Volterra, A.D. Myshkis, N.N. Krasovskii, B. Razumikhin, N. Minorsky, R. Bellman, A. Halanay, J. Hale and other mathematicians.

The problem of stability analysis of various equations continues to attract the attention of many specialists despite its long history. It is still one of the most burning problems because of the absence of its complete solution. For many years the basic method for stability analysis has been the use of Lyapunov functionals, from which many strong results have been obtained. We do not discuss this method here because it has been well covered in several excellent books. It should be noted that finding Lyapunov type functionals for vector equations is often connected with serious mathematical difficulties, especially in regard to nonautonomous equations. To the contrary, the stability conditions presented in this book are mainly formulated in terms of the determinants and eigenvalues of auxiliary matrices dependent on a parameter. This fact allows us to apply well-known results of the theory of matrices to stability analysis.

One of the methods considered in the book is the freezing method. That method was introduced by V.M. Alekseev in 1960 forstability analysis of ordinary differential equations and extended to functional differential equations by the author.

We also consider some classes of equations with causal mappings. These equations include differential, differential-delay, integro-differential and other traditional equations. The stability theory of nonlinear equations with causal mappings is in an early stage of development.

Furthermore, in 1949 M.A. Aizerman conjectured that a single input-single output system is absolutely stable in the Hurwitz angle. That hypothesis created great interest among specialists. Counter-examples were set up that demonstrated it was not, in general, true. Therefore, the following problem arose: to find the class of systems that satisfy Aizerman's hypothesis. The author has shown that any system satisfies the Aizerman hypothesis if its impulse function is non-negative. A similar result was proved for multivariable systems.

On the other hand, in 1977 A.D. Myshkis pointed out the importance of consideration of the generalized Aizerman problem for retarded systems. In 2000 it was proved by the author, that a retarded system satisfies the generalized Aizerman hypothesis if its Green function is non-negative.

3. The aim of the book is to provide new tools for specialists in the stability theory of functional differential equations, control system theory and mechanics.

This is the first book that:

i) gives a systematic exposition of an approach to stability analysis of vector differential delay equations based on estimates for matrix-valued functions allowing us to investigate various classes of equations from a unified viewpoint;
ii) contains a solution of the Aizerman–Myshkis problem;
iii) develops the Hill method for functional differential equations with periodic coefficients;
iv) presents an integral version of the generalized Bohl–Perron principle.

It also includes the freezing method for systems with delay and investigates integrally small perturbations of differential delay equations with matrix coefficients.

The book is intended not only for specialists in stability theory, but for anyone interested in various applications who has had at least a first year graduate level course in analysis.

I was very fortunate to have fruitful discussions with the late Professors M.A. Aizerman, M.A. Krasnosel'skii, A.D. Myshkis, A. Pokrovskii, and A.A. Voronov, to whom I am very grateful for their interest in my investigations.

Chapter 1

Preliminaries

1.1 Banach and Hilbert spaces

In Sections 1–3 we recall very briefly some basic notions of the theory of Banach and Hilbert spaces. More details can be found in any textbook on Banach and Hilbert spaces (e.g., [2] and [16]).

Denote the set of complex numbers by $\mathbb{C}$ and the set of real numbers by $\mathbb{R}$.

A linear space X over $\mathbb{C}$ is called *a (complex) linear normed space* if for any $x \in X$ a non-negative number $\|x\|_X = \|x\|$ is defined, called the norm of x, having the following properties:

1. $\|x\| = 0$ iff $x = 0$,
2. $\|\alpha x\| = |\alpha|\|x\|$,
3. $\|x + y\| \leq \|x\| + \|y\|$ for every $x, y \in X$, $\alpha \in \mathbb{C}$.

A sequence $\{h_n\}_{n=1}^{\infty}$ of elements of X converges *strongly* (in the norm) to $h \in X$ if

$$\lim_{n\to\infty} \|h_n - h\| = 0.$$

A sequence $\{h_n\}$ of elements of X is called the fundamental (Cauchy) one if

$$\|h_n - h_m\| \to 0 \text{ as } m, n \to \infty.$$

If any fundamental sequence converges to an element of X, then X is called *a (complex) Banach space.*

In a linear space H over $\mathbb{C}$ for all $x, y \in H$, let a number (x, y) be defined, such that

1. $(x, x) > 0$, if $x \neq 0$, and $(x, x) = 0$, if $x = 0$,
2. $(x, y) = \overline{(y, x)}$,
3. $(x_1 + x_2, y) = (x_1, y) + (x_2, y)$ $(x_1, x_2 \in H)$,
4. $(\lambda x, y) = \lambda(x, y)$ $(\lambda \in \mathbb{C})$.

Then $(.,.)$ is called the scalar product. Define in H the norm by

$$\|x\| = \sqrt{(x,x)}.$$

If H is a Banach space with respect to this norm, then it is called *a Hilbert space.* The Schwarz inequality

$$|(x,y)| \le \|x\| \, \|y\|$$

is valid.

If, in an infinite-dimensional Hilbert space, there is a countable set whose closure coincides with the space, then that space is said to be *separable.* Any separable Hilbert space H possesses an orthonormal basis. This means that there is a sequence $\{e_k \in H\}_{k=1}^{\infty}$ such that

$$(e_k, e_j) = 0 \text{ if } j \ne k \quad \text{and} \quad (e_k, e_k) = 1 \ \ (j, k = 1, 2, \dots)$$

and any $h \in H$ can be represented as

$$h = \sum_{k=1}^{\infty} c_k e_k$$

with

$$c_k = (h, e_k), \ k = 1, 2, \dots .$$

Besides the series strongly converges.

Let X and Y be Banach spaces. A function $f : X \to Y$ is continuous if for any $\epsilon > 0$, there is a $\delta > 0$, such that $\|x - y\|_X \le \delta$ implies $\|f(x) - f(y)\|_Y \le \epsilon$.

Theorem 1.1.1 (The Urysohn theorem). *Let A and B be disjoint closed sets in a Banach space X. Then there is a continuous function f defined on X such that*

$$0 \le f(x) \le 1, \ f(A) = 1 \quad \textit{and} \quad f(B) = 0.$$

For the proof see, for instance, [16, p. 15].

Let a function $x(t)$ be defined on a real segment $[0, T]$ with values in X. An element $x'(t_0)$ $(t_0 \in (0,T))$ is the derivative of $x(t)$ at t_0 if

$$\left\| \frac{x(t_0 + h) - x(t_0)}{h} - x'(t_0) \right\| \to 0 \text{ as } |h| \to 0.$$

Let $x(t)$ be continuous at each point of $[0, T]$. Then one can define the Riemann integral as the limit in the norm of the integral sums:

$$\lim_{\max |\Delta t_k^{(n)}| \to 0} \sum_{k=1}^{n} x(t_k^{(n)}) \Delta t_k^{(n)} = \int_0^T x(t)dt,$$

$$\left(0 = t_0^{(n)} < t_1^{(n)} < \dots < t_n^{(n)} = T, \Delta t_k^{(n)} = t_k^{(n)} - t_{k-1}^{(n)}\right).$$

1.2 Examples of normed spaces

The following spaces are examples of normed spaces. For more details see [16, p. 238].

1. The complex n-dimensional Euclidean space $\mathbb{C}^n$ with the norm

$$\|x\|_n = \left(\sum_{k=1}^{n} |x_k|^2\right)^{1/2} \qquad (x = \{x_k\}_{k=1}^{n} \in \mathbb{C}^n).$$

2. The space $B(S)$ is defined for an arbitrary set S and consists of all bounded scalar functions on S. The norm is given by

$$\|f\| = \sup_{s \in S} |f(s)|.$$

3. The space $C(S)$ is defined for a topological space S and consists of all bounded continuous scalar functions on S. The norm is

$$\|f\| = \sup_{s \in S} |f(s)|.$$

4. The space $L^p(S)$ is defined for any real number p, $1 \leq p < \infty$, and any set S having a finite Lebesgue measure. It consists of those measurable scalar functions on S for which the norm

$$\|f\| = \left[\int_S |f(s)|^p ds\right]^{1/p}$$

is finite.

5. The space $L^\infty(S)$ is defined for any set S having a finite Lebesgue measure. It consists of all essentially bounded measurable scalar functions on S. The norm is

$$\|f\| = \operatorname{ess\,sup}_{s \in S} |f(s)|.$$

Note that the Hilbert space has been defined by a set of abstract axioms. It is noteworthy that some of the concrete spaces defined above satisfy these axioms, and hence are special cases of abstract Hilbert space. Thus, for instance, the n-dimensional space $\mathbb{C}^n$ is a Hilbert space, if the inner product (x, y) of two elements

$$x = \{x_1, \ldots, x_n\} \quad \text{and} \quad y = \{y_1, \ldots, y_n\}$$

is defined by the formula

$$(x, y) = \sum_{k=1}^{n} x_k \overline{y}_k.$$

In the same way, complex l^2 space is a Hilbert space if the scalar product (x, y) of the vectors $x = \{x_k\}$ and $y = \{y_k\}$ is defined by the formula

$$(x, y) = \sum_{k=1}^{\infty} x_k \overline{y}_k.$$

Also the complex space $L^2(S)$ is a Hilbert space with the scalar product

$$(f,g) = \int_S f(s)\overline{g}(s)ds.$$

1.3 Linear operators

An operator A, acting from a Banach space X into a Banach space Y, is called a linear one if

$$A(\alpha x_1 + \beta x_2) = \alpha A x_1 + \beta A x_2$$

for any $x_1, x_2 \in X$ and $\alpha, \beta \in \mathbb{C}$. If there is a constant a, such that the inequality

$$\|Ah\|_Y \leq a\|h\|_X \text{ for all } h \in X$$

holds, then the operator is said to be bounded. The quantity

$$\|A\|_{X\to Y} := \sup_{h\in X} \frac{\|Ah\|_Y}{\|h\|_X}$$

is called the norm of A. If $X = Y$ we will write $\|A\|_{X\to X} = \|A\|_X$ or simply $\|A\|$.

Under the natural definitions of addition and multiplication by a scalar, and the norm, the set $B(X,Y)$ of all bounded linear operators acting from X into Y becomes a Banach space. If $Y = X$ we will write $B(X,X) = B(X)$. A sequence $\{A_n\}$ of bounded linear operators from $B(X,Y)$ converges *in the uniform operator topology* (in the operator norm) to an operator A if

$$\lim_{n\to\infty} \|A_n - A\|_{X\to Y} = 0.$$

A sequence $\{A_n\}$ of bounded linear operators *converges strongly* to an operator A if the sequence of elements $\{A_n h\}$ strongly converges to Ah for every $h \in X$.

If ϕ is a linear operator, acting from X into $\mathbb{C}$, then it is called a linear functional. It is bounded (continuous) if $\phi(x)$ is defined for any $x \in X$, and there is a constant a such that the inequality

$$\|\phi(h)\|_Y \leq a\|h\|_X \text{ for all } h \in X$$

holds. The quantity

$$\|\phi\|_X := \sup_{h\in X} \frac{|\phi(h)|}{\|h\|_X}$$

is called *the norm of the functional* ϕ. All linear bounded functionals on X form a Banach space with that norm. This space is called the space *dual* to X and is denoted by X^*.

In the sequel $I_X = I$ is the identity operator in $X : Ih = h$ for any $h \in X$.

The operator A^{-1} is the inverse one to $A \in B(X,Y)$ if $AA^{-1} = I_Y$ and $A^{-1}A = I_X$.

Let $A \in B(X,Y)$. Consider a linear bounded functional f defined on Y. Then on X the linear bounded functional $g(x) = f(Ax)$ is defined. The operator realizing the relation $f \to g$ is called the operator A^* *dual* (*adjoint*) to A. By the definition

$$(A^* f)(x) = f(Ax) \ (x \in X).$$

The operator A^* is a bounded linear operator acting from Y^* to X^*.

Theorem 1.3.1. *Let $\{A_k\}$ be a sequence of linear operators acting from a Banach space X to a Banach space Y. Let for each $h \in X$,*

$$\sup_k \|A_k h\|_Y < \infty.$$

Then the operator norms of $\{A_k\}$ are uniformly bounded. Moreover, if $\{A_n\}$ strongly converges to a (linear) operator A, then

$$\|A\|_{X \to Y} \leq \sup_n \|A_n\|_{X \to Y}.$$

For the proof see, for example, [16, p. 66].

A point λ of the complex plane is said to be a regular point of an operator A, if the operator $R_\lambda(A) := (A - I\lambda)^{-1}$ (the resolvent) exists and is bounded. The complement of all regular points of A in the complex plane is the *spectrum* of A. The spectrum of A is denoted by $\sigma(A)$.

The quantity

$$r_s(A) = \sup_{s \in \sigma(A)} |s|$$

is the *spectral radius* of A. The Gel'fand formula

$$r_s(A) = \lim_{k \to \infty} \sqrt[k]{\|A^k\|}$$

is valid. The limit always exists. Moreover,

$$r_s(A) \leq \sqrt[k]{\|A^k\|}$$

for any integer $k \geq 1$. So

$$r_s(A) \leq \|A\|.$$

If there is a nontrivial solution e of the equation $Ae = \lambda(A)e$, where $\lambda(A)$ is a number, then this number is called an eigenvalue of operator A, and $e \in H$ is an eigenvector corresponding to $\lambda(A)$. Any eigenvalue is a point of the spectrum. An eigenvalue $\lambda(A)$ has the (algebraic) multiplicity $r \leq \infty$ if

$$\dim(\cup_{k=1}^{\infty} \ker(A - \lambda(A)I)^k) = r.$$

In the sequel $\lambda_k(A)$, $k = 1, 2, \dots$ are the eigenvalues of A repeated according to their multiplicities.

A vector v satisfying $(A - \lambda(A)I)^n v = 0$ for a natural n, is a root vector of operator A corresponding to $\lambda(A)$.

An operator V is called a quasinilpotent one, if its spectrum consists of zero, only.

On a linear manifold $D(A)$ of a Banach space X,let there be defined a linear operator A, mapping $D(A)$ into a Banach space Y. Then $D(A)$ is called the domain of A. A linear operator A is called a closed operator, if from $x_n \in X \to x_0$ and $Ax_n \to y_0$ in the norm, it follows that $x_0 \in D(A)$ and $Ax_0 = y_0$.

Theorem 1.3.2 (The Closed Graph theorem). *A closed linear map defined on the all of a Banach space, and with values in a Banach space, is continuous.*

For the proof see [16, p. 57].

Theorem 1.3.3 (The Riesz–Thorin theorem). *Assume T is a bounded linear operator from $L^p(\Omega_1)$ to $L^p(\Omega_2)$ and at the same time from $L^q(\Omega_1)$ to $L^q(\Omega_2)$ $(1 \le p, q \le \infty)$. Then it is also a bounded operator from $L^r(\Omega_1)$ to $L^r(\Omega_2)$ for any r between p and q. In addition the following inequality for the norms holds:*

$$\|T\|_{L^r(\Omega_1)\to L^r(\Omega_2)} \le \max\{\|T\|_{L^p(\Omega_1)\to L^p(\Omega_2)}, \|T\|_{L^q(\Omega_1)\to L^q(\Omega_2)}\}.$$

For the proof (in a more general situation) see [16, Section VI.10.11].

Theorem 1.3.4. *Let $f \in L^1(\Omega)$ be a fixed integrable function and let T be the operator of convolution with f, i.e., for each function $g \in L^p(\Omega)$ $(p \ge 1)$ we have*

$$(Tg)(t) = \int_\Omega f(t-s)g(s)ds.$$

Then

$$\|Tg\|_{L^p(\Omega)} \le \|f\|_{L^1(\Omega)}\|g\|_{L^p(\Omega)}.$$

For the proof see [16, p. 528].

Now let us consider operators in a Hilbert space H. A bounded linear operator A^* is adjoint to A, if

$$(Af, g) = (f, A^*g) \text{ for every } h, g \in H.$$

The relation $\|A\| = \|A^*\|$ is true. A bounded operator A *is a selfadjoint* one, if $A = A^*$. A is *a unitary operator*, if $AA^* = A^*A = I$. Here and below $I \equiv I_H$ is the identity operator in H. A selfadjoint operator A is positive (negative) definite, if

$$(Ah, h) \ge 0 \ \ ((Ah, h) \le 0) \text{ for every } h \in H.$$

A selfadjoint operator A is strongly positive (strongly negative) definite, if there is a constant $c > 0$, such that

$$(Ah, h) \ge c\,(h, h) \ \ ((Ah, h) < -c\,(h, h)) \text{ for every } h \in H.$$

A bounded linear operator satisfying the relation $AA^* = A^*A$ is called *a normal operator*. It is clear that unitary and selfadjoint operators are examples of normal ones. The operator $B \equiv A^{-1}$ is the inverse one to A, if $AB = BA = I$. An operator P is called *a projection* if $P^2 = P$. If, in addition, $P^* = P$, then it is called *an orthogonal projection* (*an orthoprojection*). The spectrum of a selfadjoint operator is real, the spectrum of a unitary operator lies on the unit circle.

1.4 Ordered spaces and Banach lattices

Following [91], let us introduce an inequality relation for normed spaces which can be used analogously to the inequality relation for real numbers.

A non-empty set M with a relation $\leq$ is said to be an ordered set, whenever the following conditions are satisfied.

i) $x \leq x$ for every $x \in M$,
ii) $x \leq y$ and $y \leq x$ implies that $x = y$ and
iii) $x \leq y$ and $y \leq z$ implies that $x \leq z$.

If, in addition, for any two elements $x, y \in M$ either $x \leq y$ or $y \leq x$, then M is called a totally ordered set. Let A be a subset of an ordered set M. Then $x \in M$ is called an upper bound of A, if $y \leq x$ for every $y \in A$. $z \in M$ is called a lower bound of A, if $y \geq z$ for all $y \in A$. Moreover, if there is an upper bound of A, then A is said to be bounded from above. If there is a lower bound of A, then A is called bounded from below. If A is bounded from above and from below, then we will briefly say that A is order bounded. Let

$$[x, y] = \{z \in M : \; x \leq z \leq y\}.$$

That is, $[x, y]$ is an order interval.

An ordered set $(M, \leq)$ is called *a lattice*, if any two elements $x, y \in M$ have a least upper bound denoted by $\sup(x, y)$ and a greatest lower bound denoted by $\inf(x, y)$. Obviously, a subset A is order bounded, if and only if it is contained in some order interval.

Definition 1.4.1. A real vector space E which is also an ordered set is called an ordered vector space, if the order and the vector space structure are compatible in the following sense: if $x, y \in E$, such that $x \leq y$, then $x + z \leq y + z$ for all $z \in E$ and $ax \leq ay$ for any positive number a. If, in addition, $(E, \leq)$ is a lattice, then E is called a Riesz space (or a vector lattice).

Let E be a Riesz space. The *positive cone* E_+ of E consists of all $x \in E$, such that $x \geq 0$. For every $x \in E$ let

$$x^+ = \sup\,(x, 0), \; x^- = \inf\,(-x, 0), \; |x| = \sup\,(x, -x)$$

be the *positive part, the negative part and the absolute value of* x, respectively.

Example 1.4.2. Let $E = \mathbb{R}^n$ and

$$R^n_+ = \{(x_1, \dots, x_n) \in \mathbb{R}^n : x_k \geq 0 \text{ for all } k\}.$$

Then R^n_+ is a positive cone and for $x = (x_1, \dots, x_n), y = (y_1, \dots, y_n) \in \mathbb{R}^n$, we have

$$x \leq y \text{ iff } x_k \leq y_k \quad \text{and} \quad |x| = (|x_1|, \dots, |x_n|).$$

Example 1.4.3. Let X be a non-empty set and let $B(X)$ be the collection of all bounded real-valued functions defined on X.

It is a simple and well-known fact that $B(X)$ is a vector space ordered by the positive cone

$$B(X)_+ = \{f \in B(X) : f(t) \geq 0 \text{ for all } t \in X\}.$$

Thus $f \geq g$ holds, if and only if $f - g \in B(X)_+$. Obviously, the function $h_1 = \sup\,(f, g)$ is defined by

$$h_1(t) = \max\,\{f(t), g(t)\}$$

and the function $h_2 = \inf\,(f, g)$ is defined by

$$h_2(t) = \min\,\{f(t), g(t)\}$$

for every $t \in X$ and $f, g \in B(X)$. This shows that $B(X)$ is a Riesz space and the absolute value of f is $|f(t)|$.

Definition 1.4.4. Let E be a Riesz space furnished with a norm $\|.\|$, satisfying $\|x\| \leq \|y\|$ whenever $|x| \leq |y|$. In addition, let the space E be complete with respect to that norm. Then E is called a Banach lattice.

The norm $\|.\|$ in a Banach lattice E is said to be *order continuous*, if

$$\inf\{\|x\| : x \in A\} = 0$$

for any down-directed set $A \subset E$, such that $\inf\{x \in A\} = 0$, cf. [91, p. 86].

The real spaces $C(K), L^p(K)$ $(K \subseteq \mathbb{R}^n)$ and l^p $(p \geq 1)$ are examples of Banach lattices.

A bounded linear operator T in E is called *a positive one*, if from $x \geq 0$ it follows that $Tx \geq 0$.

1.5 The abstract Gronwall lemma

In this section E is a Banach lattice with the positive cone E_+.

Lemma 1.5.1 (The abstract Gronwall lemma). *Let T be a bounded linear positive operator acting in E and having the spectral radius*

$$r_s(T) < 1.$$

Let $x, f \in E_+$. Then the inequality

$$x \leq f + Tx$$

implies $x \leq y$ where y is a solution of the equation

$$y = f + Ty.$$

Proof. Let $Bx = f + Tx$. Then $x \leq Bx$ implies $x \leq Bx \leq B^2 x \leq \cdots \leq B^m x$. This gives

$$x \leq B^m x = \sum_{k=0}^{m-1} T^k f + T^m x \to (I - T)^{-1} f = y \text{ as } m \to \infty.$$

Since $r_s(T) < 1$, the von Neumann series converges. □

We will say that $F : E \to E$ is a *non-decreasing mapping* if $v \leq w \ \ (v, w \in E)$ implies $F(v) \leq F(w)$.

Lemma 1.5.2. *Let $F : E \to E$ be a non-decreasing mapping, and $F(0) = 0$. In addition, let there be a positive linear operator T in E, such that the conditions*

$$|F(v) - F(w)| \leq T|v - w| \ \ (v, w \in E), \tag{5.1}$$

and $r_s(T) < 1$ hold. Then the inequality

$$x \leq F(x) + f \ \ (x, f \in E_+)$$

implies that $x \leq y$ where y is a solution of the equation

$$y = F(y) + f.$$

Moreover, the inequality

$$z \geq F(z) + f \ \ (z, f \in E_+)$$

implies that $z \geq y$.

Proof. We have $x = F(x) + h$ with an $h < f$. Thanks to (5.1) and the condition $r_s(T) < 1$, the mappings $F_f := F + f$ and $F_h := F + h$ have the following properties: F_f^m and F_h^m are contracting for some integer m. So thanks to the generalized contraction mapping theorem [115], $F_f^k(f) \to x, F_h^k(f) \to y$ as $k \to \infty$. Moreover, $F_f^k(f) \geq F_h^k(f)$ for all $k = 1, 2, \ldots$, since F is non-decreasing and $h \leq f$. This proves the inequality $x \geq y$. Similarly the inequality $x \leq z$ can be proved. □

1.6 Integral inequalities

Let $C(J, \mathbb{R}^n)$ be a space of real vector-valued functions defined, bounded and continuous on a finite or infinite interval J. The inequalities are understood in the coordinate-wise sense.

To receive various solution estimates, we essentially use the following lemma.

Lemma 1.6.1. *Let $\hat{K}(t,s)$ be a matrix kernel with non-negative entries, such that the integral operator*

$$(Kx)(t) = \int_J \hat{K}(t,s)x(s)ds$$

maps $C(J, \mathbb{R}^n)$ into itself and has the spectral radius $r_s(K) < 1$. Then for any non-negative continuous vector function $v(t)$ satisfying the inequality

$$v(t) \le \int_J \hat{K}(t,s)v(s)ds + f(t)$$

where f is a non-negative continuous on J vector function, the inequality $v(t) \le u(t)$ $(t \in J)$ is valid, where $u(t)$ is a solution of the equation

$$u(t) = \int_J \hat{K}(t,s)u(s)ds + f(t).$$

Similarly, the inequality

$$v(t) \ge \int_J \hat{K}(t,s)v(s)ds + f(t)$$

implies $v(t) \ge u(t)$ $(t \in J)$.

Proof. The lemma is a particular case of the abstract Gronwall lemma. □

If $J = [a,b]$ is an arbitrary finite interval and

$$(Kx)(t) = \int_a^t \hat{K}(t,s)x(s)ds \ (t \le b),$$

and the condition

$$\sup_{t \in [a,b]} \int_a^t \|\hat{K}(t,s)\|ds < \infty$$

is fulfilled with an arbitrary matrix norm, then it is simple to show that $r_s(K) = 0$.

The same equality for the spectral radius is true, if

$$(Kx)(t) = \int_t^b \hat{K}(t,s)x(s)ds \ (t \ge a),$$

provided

$$\sup_{t \in [a,b]} \int_t^b \|\hat{K}(t,s)\|ds < \infty.$$

1.7 Generalized norms

In this section nonlinear equations are considered in a space furnished with a vector (generalized) norm introduced by L. Kantorovich [108, p. 334]. Note that a vector norm enables us to use information about equations more complete than a usual (number) norm.

Throughout this section E is a Banach lattice with a positive cone E_+ and a norm $\|.\|_E$.

Let X be an arbitrary set. Assume that in X a vector metric $M(.,.)$ is defined. That is, $M(.,.)$ maps $X \times X$ into E_+ with the usual properties: for all $x, y, z \in X$

a) $M(x, y) = 0$ iff $x = y$;
b) $M(x, y) = M(y, x)$ and
c) $M(x, y) \leq M(x, z) + M(y, z)$.

Clearly, X is a metric space with the metric $m(x, y) = \|M(x, y)\|_E$. That is, a sequence $\{x_k \in X\}$ converges to x in the metric $m(.,.)$ iff $M(x_k, x) \to 0$ as $k \to \infty$.

Lemma 1.7.1. *Let X be a space with a vector metric $M(.,.)$: $X \times X \to E_+$, and $F(x)$ map a closed set $\Phi \subseteq X$ into itself with the property*

$$M(F(x), F(y)) \leq QM(x, y) \quad (x, y \in \Phi), \tag{7.1}$$

where Q is a positive operator in E whose spectral radius $r_s(Q)$ is less than one: $r_s(Q) < 1$. Then, if X is complete in generalized metric $M(.,.)$ (or, equivalently, in metric $m(.,.)$), F has a unique fixed point $\overline{x} \in \Phi$. Moreover, that point can be found by the method of successive approximations.

Proof. Following the usual proof of the contracting mapping theorem we take an arbitrary $x_0 \in \Phi$ and define the successive approximations by the equality

$$x_k = F(x_{k-1}) \quad (k = 1, 2, \dots).$$

Hence,

$$M(x_{k+1}, x_k) = M(F(x_k), F(x_{k-1})) \leq QM(x_k, x_{k-1}) \leq \cdots \leq Q^k M(x_1, x_0).$$

For $m > k$ we thus get

$$\begin{aligned} M(x_m, x_k) &\leq M(x_m, x_{m-1}) + M(x_{m-1}, x_k) \\ &\leq \cdots \leq \sum_{j=k}^{m-1} M(x_{j+1}, x_j) \leq \sum_{j=k}^{m-1} Q^j M(x_1, x_0). \end{aligned}$$

Inasmuch as $r_s(Q) < 1$,

$$M(x_m, x_k) \leq Q^k (I - Q)^{-1} M(x_1, x_0) \to 0, \ (k \to \infty).$$

Here and below I is the unit operator in a corresponding space. Consequently, points x_k converge in the metric $M(.,.)$ to an element $\overline{x} \in \Phi$. Since

$$\lim_{k\to\infty} F(x_k) = F(\overline{x}),$$

$\overline{x}$ is the fixed point due to (2.2). Thus, the existence is proved.

To prove the uniqueness let us assume that $y \neq \overline{x}$ is a fixed point of F as well. Then by (7.1) $M(\overline{x}, y) = M(F(\overline{x}), F(y)) \leq QM(\overline{x}, y)$. Or

$$(I - Q)^{-1} M(\overline{x}, y) \leq 0.$$

But $I - Q$ is positively invertible, because $r_s(Q) < 1$. In this way, $M(\overline{x}, y) \leq 0$. This proves the result. □

Now let X be a linear space with a vector (generalized) norm $M(.)$. That is, $M(.)$ maps X into E_+ and is subject to the usual axioms: for all $x, y \in X$

$$M(x) > 0 \text{ if } x \neq 0;\ M(\lambda x) = |\lambda| M(x) \ \ (\lambda \in \mathbf{C});\ M(x+y) \leq M(x) + M(y).$$

Following [108], we shall call E *a norming lattice, and* X *a lattice-normed space.* Clearly, X with a generalized (vector) norm $M(.) : X \to E_+$ is a normed space with the norm

$$\|h\|_X = \|M(h)\|_E \ (h \in X). \tag{7.2}$$

Now the previous lemma implies

Corollary 1.7.2. *Let* X *be a space with a generalized norm* $M(.) : X \to E_+$ *and* $F(x)$ *map a closed set* $\Phi \subseteq X$ *into itself with the property*

$$M(F(x) - F(y)) \leq QM(x - y) \ \ (x, y \in \Phi),$$

where Q *is a positive operator in* E *with* $r_s(Q) < 1$. *Then, if* X *is complete in the norm defined by* (7.2), F *has a unique fixed point* $\overline{x} \in \Phi$. *Moreover, that point can be found by the method of successive approximations.*

1.8 Causal mappings

Let $X(a, b) = X([a, b]; Y)$ $(-\infty < a < b \leq \infty)$ be a normed space of functions defined on $[a, b]$ with values in a normed space Y and the unit operator I. For example $X(a, b) = C([a, b], \mathbb{C}^n)$ or $X(a, b) = L^p([a, b], \mathbb{C}^n)$.

Let P_τ $(a < \tau < b)$ be the projections defined by

$$(P_\tau w)(t) = \begin{cases} w(t) & \text{if } a \leq t \leq \tau, \\ 0 & \text{if } \tau < t \leq b \end{cases} \qquad (w \in X(a, b)),$$

and $P_a = 0$, and $P_b = I$.

Definition 1.8.1. Let F be a mapping in $X(a,b)$ having the following properties:

$$F0 \equiv 0, \tag{8.1}$$

and for all $\tau \in [a,b]$, the equality

$$P_\tau F P_\tau = P_\tau F \tag{8.2}$$

holds. Then F will be called a causal mapping (operator).

This definition is somewhat different from the definition of the causal operator suggested in [13]; in the case of linear operators our definition coincides with the one accepted in [17]. Note that, if F is defined on a closed set $\Omega \ni 0$ of $X(a,b)$, then due to the Urysohn theorem, F can be extended by zero to the whole space. Put $X(a,\tau) = P_\tau X(a,b)$. Note that, if $X(a,b) = C(a,b)$, then $P_\tau f$ is not continuous in $C(a,b)$ for an arbitrary $f \in C(a,b)$. So P_τ is defined on the whole space $C(a,b)$ but maps $C(a,b)$ into the space $B(a,b) \supset C(a,b)$, where $B(a,b)$ is the space of bounded functions. However, if $P_\tau Ff$ is continuous on $[a,\tau]$, that is $P_\tau Ff \in C(a,\tau)$ for all $\tau \in (a,b]$, and relations (8.1) and (8.2) hold, then F is causal in $C(a,b)$.

Let us point an example of a causal mapping. To this end consider in $C(0,T)$ the mapping

$$(Fw)(t) = f(t,w(t)) + \int_0^t k(t,s,w(s))ds \quad (0 \le t \le T;\ w \in C(0,T))$$

with a continuous kernel k, defined on $[0,T]^2 \times \mathbb{R}$ and a continuous function

$$f : [0,T] \times \mathbb{R} \to \mathbb{R}, \text{ satisfying } k(t,s,0) \equiv 0 \quad \text{and} \quad f(t,0) \equiv 0.$$

For each $\tau \in (0,T)$, we have

$$(P_\tau Fw)(t) = f_\tau(t,w(t)) + P_\tau \int_0^t k(t,s,w(s))ds,$$

where

$$f_\tau(t,w(t)) = \begin{cases} f(t,w(t)) & \text{if } 0 \le t \le \tau, \\ 0 & \text{if } \tau < t \le T. \end{cases}$$

Clearly,

$$f_\tau(t,w(t)) = f_\tau(t,w_\tau(t)) \quad \text{where} \quad w_\tau = P_\tau w.$$

Moreover,

$$P_\tau \int_0^t k(t,s,w(s))ds = P_\tau \int_0^t k(t,s,w_\tau(s))ds = 0, t > \tau$$

and

$$P_\tau \int_0^t k(t,s,w(s))ds = \int_0^t k(t,s,w(s))ds = \int_0^t k(t,s,w_\tau(s))ds, t \le \tau.$$

Hence it follows that the considered mapping is causal. Note that, the integral operator

$$\int_0^c k(t,s,w(s))ds$$

with a fixed positive $c \le T$ is not causal.

1.9 Compact operators in a Hilbert space

A linear operator A mapping a normed space X into a normed space Y is said to be completely continuous (compact) if it is bounded and maps each bounded set in X into a compact one in Y. The spectrum of a compact operator is either finite, or the sequence of the eigenvalues of A converges to zero, any non-zero eigenvalue has the finite multiplicity.

This section deals with completely continuous operators acting in a separable Hilbert space H. All the results presented in this section are taken from the books [2] and [65].

Any normal compact operator can be represented in the form

$$A = \sum_{k=1}^{\infty} \lambda_k(A) E_k,$$

where E_k are eigenprojections of A, i.e., the projections defined by $E_k h = (h, d_k)d_k$ for all $h \in H$. Here d_k are the normal eigenvectors of A. Recall that eigenvectors of normal operators are mutually orthogonal.

A completely continuous quasinilpotent operator sometimes is called a Volterra operator.

Let $\{e_k\}$ be an orthogonal normal basis in H, and the series

$$\sum_{k=1}^{\infty} (Ae_k, e_k)$$

converges. Then the sum of this series is called *the trace of A:*

$$\text{Trace}\, A = \text{Tr}\ A = \sum_{k=1}^{\infty} (Ae_k, e_k).$$

An operator A satisfying the condition

$$\text{Tr}\ (A^* A)^{1/2} < \infty$$

is called *a nuclear operator.* An operator A, satisfying the relation

$$\mathrm{Tr}\,(A^*A) < \infty$$

is said to be *a Hilbert–Schmidt operator.*

The eigenvalues $\lambda_k((A^*A)^{1/2})$ $(k = 1, 2, \ldots)$ of the operator $(A^*A)^{1/2}$ are called *the singular numbers* (s-numbers) of A and are denoted by $s_k(A)$. That is,

$$s_k(A) := \lambda_k((A^*A)^{1/2}) \ \ (k = 1, 2, \ldots).$$

Enumerate singular numbers of A taking into account their multiplicity and in decreasing order. The set of completely continuous operators acting in a Hilbert space and satisfying the condition

$$N_p(A) := \left[\sum_{k=1}^{\infty} s_k^p(A)\right]^{1/p} < \infty,$$

for some $p \geq 1$, is called *the von Schatten–von Neumann ideal and is denoted by SN_p. $N_p(.)$ is called the norm of the ideal SN_p.* It is not hard to show that

$$N_p(A) = \sqrt[p]{\mathrm{Tr}\,(AA^*)^{p/2}}.$$

Thus, SN_1 is the ideal of nuclear operators (*the Trace class*) and SN_2 is the ideal of Hilbert–Schmidt operators. $N_2(A)$ is called the *Hilbert–Schmidt norm.* Sometimes we will omit index 2 of the Hilbert–Schmidt norm, i.e.,

$$N(A) := N_2(A) = \sqrt{\mathrm{Tr}\,(A^*A)}.$$

For any orthonormal basis $\{e_k\}$ we can write

$$N_2(A) = \left(\sum_{k=1}^{\infty} \|Ae_k\|^2\right)^{1/2}.$$

This equality is equivalent to the following one:

$$N_2(A) = \left(\sum_{j,k=1}^{\infty} |a_{jk}|^2\right)^{1/2},$$

where $a_{jk} = (Ae_k, e_j)$ $(j, k = 1, 2, \ldots)$ are entries of a Hilbert–Schmidt operator A in an orthonormal basis $\{e_k\}$.

For all finite $p \geq 1$, the following propositions are true (the proofs can be found in the book [65, Section 3.7]).

If $A \in SN_p$, then also $A^* \in SN_p$. If $A \in SN_p$ and B is a bounded linear operator, then both AB and BA belong to SN_p. Moreover,

$$N_p(AB) \le N_p(A)\|B\| \quad \text{and} \quad N_p(BA) \le N_p(A)\|B\|.$$

In addition, the inequality

$$\sum_{j=1}^{n} |\lambda_j(A)|^p \le \sum_{j=1}^{n} s_j^p(A) \quad (n = 1, 2, \dots)$$

is valid, cf. [65, Theorem II.3.1].

Lemma 1.9.1. *If $A \in SN_p$ and $B \in SN_q$ $(1 < p, q < \infty)$, then $AB \in SN_s$ with*

$$\frac{1}{s} = \frac{1}{p} + \frac{1}{q}.$$

Moreover,

$$N_s(AB) \le N_p(A)N_q(B).$$

For the proof of this lemma see [65, Section III.7]. Recall also the following result.

Theorem 1.9.2 (Lidskij's theorem). *Let $A \in SN_1$. Then*

$$\text{Tr } A = \sum_{k=1}^{\infty} \lambda_k(A).$$

The proof of this theorem can be found in [65, Section III.8].

1.10 Regularized determinants

The regularized determinant of $I - A$ with $A \in SN_p$ $(p = 1, 2, \dots)$ is defined as

$$\det_p(I - A) := \prod_{j=1}^{\infty} E_p(\lambda_j(A)),$$

where $\lambda_j(A)$ are the eigenvalues of A with their multiplicities arranged in decreasing order, and

$$E_p(z) := (1 - z)\exp\left[\sum_{m=1}^{p-1} \frac{z^m(A)}{m}\right], \; p > 1 \quad \text{and} \quad E_1(z) := 1 - z.$$

As shown below, regularized determinant are useful for the investigation of periodic systems.

The following lemma is proved in [57].

Lemma 1.10.1. *The inequality*

$$|E_p(z)| \leq \exp[\zeta_p |z|^p]$$

is valid, where

$$\zeta_p = \frac{p-1}{p} \quad (p \neq 1, p \neq 3) \quad \textit{and} \quad \zeta_1 = \zeta_3 = 1.$$

From this lemma one can immediately obtain the following result.

Lemma 1.10.2. *Let $A \in SN_p$ $(p = 1, 2, \ldots)$. Then*

$$|\det_p(I - A)| \leq \exp[\zeta_p N_p^p(A)].$$

Let us point out also the lower bound for regularized determinants which has been established in [45]. To this end denote by L *a Jordan contour connecting* 0 *and* 1, *lying in the disc* $\{z \in \mathbb{C} : |z| \leq 1\}$ *and not containing the points* $1/\lambda_j$ for any eigenvalue λ_j of A, such that

$$\phi_L(A) := \inf_{s \in L;\ k=1,2,\ldots} |1 - s\lambda_k| > 0.$$

Theorem 1.10.3. *Let $A \in SN_p$ for an integer $p \geq 1$ and $1 \notin \sigma(A)$. Then*

$$\left|\det_p(I - A)\right| \geq \exp\left[-\frac{\zeta_p N_p^p(A)}{\phi_L(A)}\right].$$

1.11 Perturbations of determinants

Now let us consider perturbations of determinants. Let X and Y be complex normed spaces with norms $\|.\|_X$ and $\|.\|_Y$, respectively, and F be a Y-valued function defined on X. Assume that $F(C + \lambda\tilde{C})$ $(\lambda \in \mathbb{C})$ is an entire function for all $C, \tilde{C} \in X$. That is, for any $\phi \in Y^*$, the functional $< \phi, F(C+\lambda\tilde{C}) >$ defined on Y is an entire scalar-valued function of λ. In [44], the following lemma has been proved.

Lemma 1.11.1. *Let $F(C + \lambda\tilde{C})$ $(\lambda \in \mathbb{C})$ be an entire function for all $C, \tilde{C} \in X$ and there be a monotone non-decreasing function $G : [0, \infty) \to [0, \infty)$, such that $\|F(C)\|_Y \leq G(\|C\|_X)$ $(C \in X)$. Then*

$$\|F(C) - F(\tilde{C})\|_Y \leq \|C - \tilde{C}\|_X \, G\left(1 + \frac{1}{2}\|C + \tilde{C}\|_X + \frac{1}{2}\|C - \tilde{C}\|_X\right) \quad (C, \tilde{C} \in X).$$

Lemmas 1.10.1 and 1.11.1 imply the following result.

Corollary 1.11.2. *The inequality*

$$\left|\det_p (I - A) - \det_p (I - B)\right| \le \delta_p(A, B) \quad (A, B \in SN_p, 1 \le p < \infty)$$

is true, where

$$\delta_p(A, B) := N_p(A - B)\ \exp\left[\zeta_p\ (1 + \frac{1}{2}(N_p(A + B) + N_p(A - B)))^p\right]$$
$$\le N_p(A - B)\ \exp\left[(1 + N_p(A) + N_p(B)))^p\right].$$

Now let A and B be $n \times n$-matrices. Then due to the inequality between the arithmetic and geometric mean values,

$$|\det\ A|^2 = \prod_{k=1}^{n} |\lambda_k(A)|^2 \le \left(\frac{1}{n}\sum_{k=1}^{n} |\lambda_k(A)|^2\right)^{1/n}.$$

Thus,

$$|\det\ A| \le \frac{1}{n^{n/2}} N_2^n(A).$$

Moreover, $|\det\ A| \le \|A\|^n$ for an arbitrary matrix norm. Hence, Lemma 1.11.1 implies our next result.

Corollary 1.11.3. *Let A and B be $n \times n$-matrices. Then*

$$|\det\ A - \det\ B| \le \frac{1}{n^{n/2}} N_2(A - B)\left[1 + \frac{1}{2}N_2(A - B) + \frac{1}{2}N_2(A + B)\right]^n$$

and

$$|\det\ A - \det\ B| \le \|A - B\|\left[1 + \frac{1}{2}\|A - B\| + \frac{1}{2}\|A - B\|\right]^n$$

for an arbitrary matrix norm $\|.\|$.

Now let us recall the well-known inequality for determinants.

Theorem 1.11.4 (Ostrowski [97]). *Let $A = (a_{jk})$ be a real $n \times n$-matrix. Then the inequality*

$$|\det\ A| \ge \prod_{j=1}^{n}\left(|a_{jj}| - \sum_{m=1, m\ne j}^{n} |a_{jm}|\right)$$

is valid, provided

$$|a_{jj}| > \sum_{m=1, m\ne j}^{n} |a_{jm}| \quad (j = 1, \dots, n).$$

1.12 Matrix functions of bounded variations

A scalar function $g : [a, b] \to \mathbb{R}$ is *a function of bounded variation* if

$$\operatorname{var}(g) = \operatorname{Var}_{t\in[a,b]}\, g(t) := \sup_P \sum_{i=0}^{n-1} |g(t_{i+1}) - g(t_i)| < \infty,$$

where the supremum is taken over the set of all partitions P of the interval $[a, b]$.

Any function of bounded variation $g : [a, b] \to \mathbb{R}$ is a difference of bounded nondecreasing functions. If g is differentiable and its derivative is integrable then its variation satisfies

$$\operatorname{var}(g) \le \int_a^b |g'(s)|ds.$$

For more details see [16, p. 140]. Sometimes we will write

$$\operatorname{var}(g) = \int_a^b |dg(s)|.$$

Let $\|x\|_n$ be the Euclidean norm of a vector x and $\|A\|_n$ be the spectral norm of a matrix A. The norm of f in $C([a, b], \mathbb{C}^n)$ is $\sup_t \|f(t)\|_n$, in $L^p([a, b], \mathbb{C}^n)$ $(1 \le p < \infty)$ its norm is $(\int_a^b \|f(t)\|_n^p)^{1/p}$, in $L^\infty([a, b], \mathbb{C}^n)$ its norm is $\operatorname{vrai\,sup}_t \|f(t)\|_n$.

For a real matrix-valued function $R_0(s) = (r_{ij}(s))_{i,j=1}^n$ defined on a real finite segment $[a, b]$, whose entries have bounded variations

$$\operatorname{var}(r_{ij}) = \operatorname{var}_{s\in[a,b]}\, r_{ij}(s),$$

we can define its variation as the matrix

$$\operatorname{Var}(R_0) = \operatorname{Var}_{s\in[a,b]}\, R_0(s) = (\operatorname{var}(r_{ij}))_{i,j=1}^n.$$

For a $c > b$ and an $f \in C([a, c], \mathbb{C}^n)$ put

$$E_0 f(t) = \int_a^b dR_0(s) f(t - s) \quad (b \le t \le c).$$

Below we prove that E_0 is bounded in the norm of L^p $(p \ge 1)$ on the set of continuous functions and therefore can be extended to the whole space L^p, since that set is dense in L^p.

Let

$$\operatorname{var}(R_0) := \|\operatorname{Var}(R_0)\|_n,$$

and

$$\zeta_1(R_0) := \sum_{j=1}^n \sqrt{\sum_{k=1}^n (\operatorname{var}(r_{jk}))^2}.$$

So $\operatorname{var}(R_0)$ is the spectral norm of matrix $\operatorname{Var}(R_0)$.

Lemma 1.12.1. *Suppose all the entries r_{jk} of the matrix function R_0 defined on $[a,b]$ have bounded variations. Then the inequalities*

$$\|E_0\|_{C([a,c],\mathbb{C}^n)\to C([b,c],\mathbb{C}^n)} \le \sqrt{n}\ \operatorname{var}(R_0), \tag{12.1}$$

$$\|E_0\|_{L^\infty([a,c],\mathbb{C}^n)\to L^\infty([b,c],\mathbb{C}^n)} \le \sqrt{n}\ \operatorname{var}(R_0), \tag{12.2}$$

$$\|E_0\|_{L^2([a,c],\mathbb{C}^n)\to L^2([b,c],\mathbb{C}^n)} \le \operatorname{var}(R_0), \tag{12.3}$$

and

$$\|E_0\|_{L^1([a,c],\mathbb{C}^n)\to L^1([b,c],\mathbb{C}^n)} \le \zeta_1\,(R_0) \tag{12.4}$$

are valid.

Proof. Let $f(t) = (f_k(t))_{k=1}^n \in C([a,c],\mathbb{C}^n)$. For each coordinate $(E_0 f)_j(t)$ of $E_0 f(t)$ we have

$$|(E_0 f)_j(t)| = \left|\int_a^b \sum_{k=1}^n f_k(t-s)dr_{jk}(s)\right|$$
$$\le \sum_{k=1}^n \int_a^b |dr_{jk}| \max_{a\le s\le b} |f_k(t-s)| = \sum_{k=1}^n \operatorname{var}(r_{jk}) \max_{a\le s\le b} |f_k(t-s)|.$$

Hence,

$$\sum_{j=1}^n |(E_0 f)_j(t)|^2 \le \sum_{j=1}^n \left(\sum_{k=1}^n \operatorname{var}(r_{jk})\|f_k\|_{C(a,c)}\right)^2$$
$$= \|\operatorname{Var}(R_0)\ \nu_C\|_n^2 \le (\operatorname{var}(R_0)\|\nu_C\|_n)^2 \quad (b\le t\le c),$$

where $\nu_C = (\|f_k\|_{C(a,c)})_{k=1}^n$, $\|.\|_{C(a,c)} = \|.\|_{C([a,c],\mathbb{C})}$.

But

$$\|\nu_C\|_n^2 = \sum_{k=1}^n \|f_k\|^2_{C(a,c)} \le n \max_k \|f_k\|^2_{C(a,c)} \le n \sup_t \sum_{k=1}^n \|f_k(t)\|_n^2 = n\|f\|^2_{C([a,c],\mathbb{C}^n)}.$$

So

$$\|E_0 f\|_{C([b,c],\mathbb{C}^n)} \le \sqrt{n}\operatorname{var}(R_0)\|f\|_{C([a,c],\mathbb{C}^n)}$$

and thus inequality (12.1) is proved.

In the case of the space L^∞ by inequality (12.1) we have

$$\|E_0 f\|_{L^\infty([b,c],\mathbb{C}^n)} \le \sqrt{n}\ \operatorname{var}(R_0)\|f\|_{L^\infty([a,c],\mathbb{C}^n)}$$

for a continuous function f. But the set of continuous functions is dense in L^∞. So the previous inequality is valid on the whole space L^∞. This proves (12.2).

Now consider the norm in space L^2. We have

$$\int_b^c |(E_0 f)_j(t)|^2 dt \le \int_b^c \left(\sum_{k=1}^n \int_a^b |f_k(t-s)||dr_{jk}(s)| \right)^2 dt$$
$$= \int_a^b \int_a^b \sum_{i=1}^n \sum_{k=1}^n |dr_{jk}(s)||dr_{ji}(s_1)| \int_b^c |f_k(t-s) f_i(t-s_1)| dt.$$

By the Schwarz inequality

$$\left(\int_b^c |f_k(t-s) f_i(t-s_1)| dt \right)^2 \le \int_b^c |f_k(t-s)|^2 dt \int_b^c |f_i(t-s_1)|^2 dt$$
$$\le \int_a^c |f_k(t)|^2 dt \int_a^c |f_i(t)|^2 dt.$$

Thus

$$\int_b^c |(E_0 f)_j(t)|^2 dt \le \sum_{i=1}^n \sum_{k=1}^n \operatorname{var}(r_{jk}) \operatorname{var}(r_{ji}) \|f_k\|_{L^2(a,c)} \|f_i\|_{L^2(a,c)}$$
$$= \left(\sum_{k=1}^n \operatorname{var}(r_{jk}) \|f_k\|_{L^2(a,c)} \right)^2 \quad (\|f_k\|_{L^2(a,c)} = \|f_k\|_{L^2([a,c],\mathbb{C})})$$

and therefore

$$\sum_{j=1}^n \int_b^c |(E_0 f)_j(t)|^2 dt \le \sum_{j=1}^n \left(\sum_{k=1}^n \operatorname{var}(r_{jk}) \|f_k\|_{L^2(a,c)} \right)^2$$
$$= \| \operatorname{Var}(R_0)\, \nu_2 \|_n^2 \le (\operatorname{var}(R_0) \|\nu_2\|_n)^2$$

where ν_2 is the vector with the coordinates $\|f_k\|_{L^2(a,c)}$. But $\|\nu_2\|_n = \|f\|_{L^2([a,c],\mathbb{C}^n)}$. So (12.3) is also proved.

Similarly, for an $f(t) = (f_k(t))_{k=1}^n \in L^1([a,c],\mathbb{C}^n)$ we obtain

$$\int_b^c |(E_0 f)_j(t)| dt \le \sum_{k=1}^n \int_a^b \int_b^c |f_k(t-s)| dt |dr_{jk}(s)| \le \sum_{k=1}^n \operatorname{var}(r_{jk}) \int_a^c |f_k(t)| dt.$$

So

$$\|E_0 f\|_{L^1([b,c],\mathbb{C}^n)} = \int_b^c \sqrt{\sum_{j=1}^n |(E_0 f)_j(t)|^2 dt} \le \int_b^c \sum_{j=1}^n |(E_0 f)_j(t)| dt$$
$$\le \sum_{j=1}^n \int_a^c \sum_{k=1}^n \operatorname{var}(r_{jk}) |f_k(t)| dt.$$

Consequently, by the Schwarz inequality

$$\|E_0 f\|_{L^1([b,c],\mathbb{C}^n)} \le \sum_{j=1}^{n} \int_a^c \sqrt{\sum_{k=1}^{n} (\mathrm{var}(r_{jk}))^2 \sum_{k=1}^{n} |f_k(t)|^2} \; dt.$$

Hence (12.4) follows, as claimed. □

The Riesz–Thorin theorem (see Section 1.3) and previous lemma imply the following result.

Corollary 1.12.2. *The inequalities*

$$\|E_0\|_{L^p([a,c],\mathbb{C}^n)\to L^p([b,c],\mathbb{C}^n)} \le \sqrt{n}\ \mathrm{var}(R_0) \ \ (c > b;\ p \ge 2)$$

and

$$\|E_0\|_{L^p([a,c],\mathbb{C}^n)\to L^p([b,c],\mathbb{C}^n)} \le \max\{\zeta_1(R_0), \sqrt{n}\ \mathrm{var}(R_0)\} \ \ (c > b;\ p \ge 1)$$

are valid.

Let us consider the operator

$$Ef(t) = \int_a^b d_s R(t,s) f(t-s) \ \ (f \in C([a,c],\mathbb{C}^n);\ b \le t \le c) \tag{12.5}$$

where $R(t,s) = (r_{ij}(t,s))_{i,j=1}^n$ is a real $n \times n$-matrix-valued function defined on $[b,c] \times [a,b]$, which is piece-wise continuous in t for each τ and

$$v_{jk} := \sup_{b \le t \le c} \mathrm{var}(r_{jk}(t,.)) < \infty \ \ (j,k = 1,\dots,n). \tag{12.6}$$

Below we prove that E is also bounded in the norm of L^p $(p \ge 1)$ on the set of continuous functions and therefore can be extended to the whole space L^p.

Let

$$Z(R) = (v_{jk})_{j,k=1}^n.$$

Since for $b \le t \le c$,

$$\left| \int_a^b f_k(t-s) dr_{jk}(t,s) \right| \le \max_{a \le s \le b} |f_k(t-s)| \int_a^b |r_{jk}(t,s)| \le v_{jk} \|f_k\|_{C(a,c)},$$

for each coordinate $(Ef)_j(t)$ of $Ef(t)$ we have

$$|(Ef)_j(t)| = \left| \int_a^b \sum_{k=1}^n f_k(t-s) dr_{jk}(t,s) \right| \le \sum_{k=1}^n v_{jk} \sup_{a \le s \le b} |f_k(t-s)|.$$

Hence,

$$\sum_{j=1}^n |(Ef)_j(t)|^2 \le \sum_{j=1}^n \left(\sum_{k=1}^n (v_{jk})\right) \|f_k\|^2_{C(a,c)} = \|Z(R)\nu_C\|_n^2 \le \|Z(R)\|_n \|\nu_C\|_n^2,$$

where η_C is the same as in the proof of the previous lemma. As shown above, $\|\nu_C\|_n^2 \le n\|f\|^2_{C([a,c],\mathbb{C}^n)}$. Thus,

$$\|Ef\|_{C([b,c],\mathbb{C}^n)} \le \sqrt{n}\|Z(R)\|_n \|f\|_{C([a,c],\mathbb{C}^n)}. \tag{12.7}$$

Moreover,

$$\|Ef\|_{L^\infty([b,c],\mathbb{C}^n)} \le \sqrt{n}\|Z(R)\|_n \|f\|_{L^\infty([a,c],\mathbb{C}^n)} \tag{12.8}$$

for a continuous function f. But the set of continuous functions is dense in L^∞. So the previous inequality is valid on the whole space. Repeating the arguments of the proof of the previous lemma we obtain

$$\|Ef\|_{L^2([b,c],\mathbb{C}^n)} \le \|Z(R)\|_n \|f\|_{L^2([a,c],\mathbb{C}^n)}. \tag{12.9}$$

Now let $f(t) = (f_k(t)) \in L^1([a,c],\mathbb{C}^n)$. Then

$$\int_b^c |(Ef)_j(t)|dt \le \sum_{k=1}^n \int_a^b \int_b^c |f_k(t-s)|dt|dr_{jk}(t,s)| \le \sum_{k=1}^n v_{jk} \int_a^c |f_k(t)|dt.$$

So

$$\|Ef\|_{L^1([b,c],\mathbb{C}^n)} = \int_b^c \sqrt{\sum_{j=1}^n |(Ef)_j(t)|^2 dt} \le \int_b^c \sum_{j=1}^n |(Ef)_j(t)|dt$$

$$\le \sum_{j=1}^n \int_a^c \sum_{k=1}^n v_{jk}|f_k(t)|dt \le \sum_{j=1}^n \int_a^c \sqrt{\sum_{k=1}^n v_{jk}^2 \sum_{k=1}^n |f_k(t)|^2}\ dt$$

Hence

$$\|Ef\|_{L^1([b,c],\mathbb{C}^n)} \le V_1(R)\|f\|_{L^1([a,c],\mathbb{C}^n)}, \tag{12.10}$$

where

$$V_1(R) = \sum_{j=1}^n \sqrt{\sum_{k=1}^n v_{jk}^2}.$$

We thus have proved the following result.

Lemma 1.12.3. *Suppose the entries $r_{jk}(t,s)$ of the matrix function $R(t,s)$ satisfy condition* (12.6). *Then the operator E defined by* (12.5) *is subject to the inequalities* (12.7)–(12.10).

Now the Riesz–Thorin theorem and the previous lemma imply the following result.

Corollary 1.12.4. *Let condition* (12.6) *hold. Then for the operator E defined by* (12.5)*, the inequalities*

$$\|E\|_{L^p([a,c],\mathbb{C}^n)\to L^p([b,c],\mathbb{C}^n)} \leq \sqrt{n}\|Z\|_n \quad (c > b;\ p \geq 2)$$

and

$$\|E\|_{L^p([a,c],\mathbb{C}^n)\to L^p([b,c],\mathbb{C}^n)} \leq V(R) \quad (c > b;\ p \geq 1)$$

are true, where $V(R) = \max\{V_1(R), \sqrt{n}\|Z(R)\|_n\}$.

1.13 Comments

The chapter contains mostly well-known results. This book presupposes a knowledge of basic operator theory, for which there are good introductory texts. The books [2] and [16] are classical. In Sections 1.5 and 1.6 we followed Sections I.9 and III.2 of the book [14]. The material of Sections 1.10 and 1.11 is adapted from the papers [57, 44] and [45]. The relevant results on regularized determinants can be found in [64]. Lemmas 1.12.1 and 11.12.3 are probably new.

Chapter 2

Some Results of the Matrix Theory

This chapter is devoted to norm estimates for matrix-valued functions, in particular, for resolvents. These estimates will be applied in the rest of the book chapters.

In Section 2.1 we introduce the notation used in this chapter. In Section 2.2 we recall the well-known representations of matrix-valued functions. In Sections 2.3 and 2.4 we collect inequalities for the resolvent and present some results on spectrum perturbations of matrices. Sections 2.5 and 2.6 are devoted to matrix functions regular on simply-connected domains containing the spectrum. Section 2.7 deals with functions of matrices having geometrically simple eigenvalues; i.e., so-called diagonalizable matrices. In the rest of the chapter we consider particular cases of functions and matrices.

2.1 Notations

Everywhere in this chapter $\|x\|$ is the Euclidean norm of $x \in \mathbb{C}^n$: $\|x\| = \sqrt{(x,x)}$ with a scalar product $(.,.) = (.,.)_{C^n}$, I is the unit matrix.

For a linear operator A in $\mathbb{C}^n$ (matrix), $\lambda_k = \lambda_k(A)$ $(k = 1, \dots, n)$ are the eigenvalues of A enumerated in an arbitrary order with their multiplicities, $\sigma(A)$ denotes the spectrum of A, A^* is the adjoint to A, and A^{-1} is the inverse to A; $R_\lambda(A) = (A - \lambda I)^{-1}$ $(\lambda \in \mathbb{C}, \lambda \notin \sigma(A))$ is the resolvent, $r_s(A)$ is the spectral radius, $\|A\| = \sup_{x \in \mathbb{C}^n} \|Ax\|/\|x\|$ is the (operator) spectral norm, $N_2(A)$ is the Hilbert–Schmidt (Frobenius) norm of A: $N_2^2(A) = \text{Trace}\, AA^*$, $A_I = (A - A^*)/2i$ is the imaginary component, $A_R = (A + A^*)/2$ is the real component,

$$\rho(A, \lambda) = \min_{k=1,\dots,n} |\lambda - \lambda_k(A)|$$

is the distance between $\sigma(A)$ and a point $\lambda \in \mathbb{C}$; $\rho(A, C)$ is the Hausdorff distance between a contour C and $\sigma(A)$. $\text{co}(A)$ denotes the closed convex hull of $\sigma(A)$,

$\alpha(A) = \max_k \operatorname{Re} \lambda_k(A)$, $\beta(A) = \min_k \operatorname{Re} \lambda_k(A)$; $r_l(A)$ is the lower spectral radius:

$$r_l(A) = \min_{k=1,\ldots,n} |\lambda_k(A)|.$$

In addition, $\mathbb{C}^{n\times n}$ is the set of complex $n \times n$-matrices.

The following quantity plays an essential role in the sequel:

$$g(A) = \left(N_2^2(A) - \sum_{k=1}^{n} |\lambda_k(A)|^2 \right)^{1/2}.$$

It is not hard to check that

$$g^2(A) \leq N_2^2(A) - |\operatorname{Trace} A^2|.$$

In Section 2.2 of the book [31] it is proved that

$$g^2(A) \leq 2N_2^2(A_I) \tag{1.1}$$

and

$$g(e^{i\tau} A + zI) = g(A) \tag{1.2}$$

for all $\tau \in \mathbb{R}$ and $z \in \mathbb{C}$.

2.2 Representations of matrix functions

2.2.1 Classical representations

In this subsection we recall some classical representations of functions of matrices. For details see [74, Chapter 6] and [9].

Let $A \in \mathbb{C}^{n\times n}$ and $M \supset \sigma(A)$ be an open simply-connected set whose boundary C consists of a finite number of rectifiable Jordan curves, oriented in the positive sense customary in the theory of complex variables. Suppose that $M \cup C$ is contained in the domain of analyticity of a scalar-valued function f. Then $f(A)$ can be defined by the generalized integral formula of Cauchy

$$f(A) = -\frac{1}{2\pi i} \int_C f(\lambda) R_\lambda(A) d\lambda. \tag{2.1}$$

If an analytic function $f(\lambda)$ is represented by the Taylor series

$$f(\lambda) = \sum_{k=0}^{\infty} c_k \lambda^k \quad \left(|\lambda| < \frac{1}{\overline{\lim}_{k\to\infty} \sqrt[k]{|c_k|}} \right),$$

then one can define $f(A)$ as

$$f(A) = \sum_{k=0}^{\infty} c_k A^k$$

provided the spectral radius $r_s(A)$ of A satisfies the inequality

$$r_s(A) \lim_{k\to\infty} \sqrt[k]{|c_k|} < 1.$$

In particular, for any matrix A,

$$e^A = \sum_{k=0}^{\infty} \frac{A^k}{k!}.$$

Consider the $n \times n$-Jordan block:

$$J_n(\lambda_0) = \begin{pmatrix} \lambda_0 & 1 & 0 & \dots & 0 \\ 0 & \lambda_0 & 1 & \dots & 0 \\ . & . & . & \dots & . \\ . & . & . & \dots & . \\ . & . & . & \dots & . \\ 0 & 0 & \dots & \lambda_0 & 1 \\ 0 & 0 & \dots & 0 & \lambda_0 \end{pmatrix},$$

then

$$f(J_n(\lambda_0)) = \begin{pmatrix} f(\lambda_0) & \frac{f'(\lambda_0)}{1!} & \dots & \frac{f^{(n-1)}(\lambda_0)}{(n-1)!} \\ 0 & f(\lambda_0) & \dots & \\ . & . & \dots & . \\ . & . & \dots & . \\ . & . & \dots & . \\ 0 & \dots & f(\lambda_0) & \frac{f'(\lambda_0)}{1!} \\ 0 & \dots & 0 & f(\lambda_0) \end{pmatrix}.$$

Thus, if A has the Jordan block-diagonal form

$$A = \operatorname{diag}(J_{m_1}(\lambda_1), J_{m_2}(\lambda_2), \dots, J_{m_{n_0}}(\lambda_{n_0})),$$

where λ_k, $k = 1, \dots, n_0$ are the eigenvalues whose geometric multiplicities are m_k, then

$$f(A) = \operatorname{diag}(f(J_{m_1}(\lambda_1)), f(J_{m_2}(\lambda_2)), \dots, f(J_{m_{n_0}}(\lambda_{n_0}))). \tag{2.2}$$

Note that in (2.2) we do not require that f is regular on a neighborhood of an open set containing $\sigma(A)$; one can take an arbitrary function which has at each λ_k derivatives up to $m_k - 1$-order.

In particular, if an $n \times n$-matrix A is diagonalizable, that is its eigenvalues have the geometric multiplicities $m_k \equiv 1$, then

$$f(A) = \sum_{k=1}^{n} f(\lambda_k) Q_k, \tag{2.3}$$

where Q_k are the eigenprojections. In the case (2.3) it is required only that f is defined on the spectrum.

Now let

$$\sigma(A) = \cup_{k=1}^{m} \sigma_k(A) \ \ (m \leq n)$$

and $\sigma_k(A) \subset M_k$ $(k = 1, \dots, m)$, where M_k are open disjoint simply-connected sets: $M_k \cap M_j = \emptyset$ $(j \neq k)$. Let f_k be regular on M_k. Introduce on $M = \cup_{k=1}^{m} M_k$ the piece-wise analytic function by $f(z) = f_j(z)$ $(z \in M_j)$. Then

$$f(A) = -\frac{1}{2\pi i} \sum_{j=1}^{m} \int_{C_j} f(\lambda) R_\lambda(A) d\lambda, \tag{2.4}$$

where $C_j \subset M_j$ are closed smooth contours surrounding $\sigma(A_j)$ and the integration is performed in the positive direction. For more details about representation (2.4) see [110, p. 49].

For instance, let M_1 and M_2 be two disjoint disks, and

$$f(z) = \begin{cases} \sin z & \text{if } z \in M_1, \\ \cos z & \text{if } z \in M_2. \end{cases}$$

Then (2.4) holds with $m = 2$.

2.2.2 Multiplicative representations of the resolvent

In this subsection we present multiplicative representations of the resolvent which lead to new representations of matrix functions.

Let $A \in \mathbb{C}^{n \times n}$ and λ_k be its eigenvalues with the multiplicities taken into account.

As it is well known, there is an orthogonal normal basis (the Schur basis) $\{e_k\}$ in which A is represented by a triangular matrix. Moreover there is the (maximal) chain P_k $(k = 1, \dots, n)$ of the invariant orthogonal projections of A. That is, $AP_k = P_k A P_k$ $(k = 1, \dots, n)$ and

$$0 = P_0 \mathbb{C}^n \subset P_1 \mathbb{C}^n \subset \cdots \subset P_n \mathbb{C}^n = \mathbb{C}^n.$$

So $\dim(P_k - P_{k-1})\mathbb{C}^n = 1$. Besides,

$$A = D + V \ \ (\sigma(A) = \sigma(D)), \tag{2.5}$$

where

$$D = \sum_{k=1}^{n} \lambda_k \Delta P_k \ \ (\Delta P_k = P_k - P_{k-1}) \tag{2.6}$$

is the diagonal part of A and V is the nilpotent part of A. That is, V is a nilpotent matrix, such that

$$VP_k = P_{k-1} A P_k \ (k = 2, \dots, n). \tag{2.7}$$

For more details see for instance, [18]. The representation (2.5) will be called the triangular (Schur) representation.

Furthermore, for $X_1, X_2, \dots, X_j \in \mathbb{C}^{n\times n}$ let

$$\prod_{1\le k\le j}^{\rightarrow} X_k \equiv X_1 X_2 \dots X_n.$$

That is, the arrow over the symbol of the product means that the indexes of the co-factors increase from left to right.

Theorem 2.2.1. *Let D and V be the diagonal and nilpotent parts of an $A \in \mathbb{C}^{n\times n}$, respectively. Then*

$$R_\lambda(A) = R_\lambda(D) \prod_{2\le k\le n}^{\rightarrow} \left[I + \frac{V\Delta P_k}{\lambda - \lambda_k} \right] \quad (\lambda \notin \sigma(A)),$$

where P_k, $k = 1, \dots, n$, is the maximal chain of the invariant projections of A.

For the proof see [31, Theorem 2.9.1]. Since

$$R_\lambda(D) = \sum_{j=1}^{n} \frac{\Delta P_k}{\lambda_j - \lambda},$$

from the previous theorem we have the following result.

Corollary 2.2.2. *The equality*

$$R_\lambda(A) = \sum_{j=1}^{n} \frac{\Delta P_j}{\lambda_j - \lambda} \prod_{j+1\le k\le n+1}^{\rightarrow} \left[I + \frac{V\Delta P_k}{\lambda - \lambda_k} \right] \quad (\lambda \notin \sigma(A))$$

is true with $V\Delta P_{n+1} = 0$.

Now (2.1) implies the representation

$$f(A) = -\frac{1}{2\pi i} \int_C f(\lambda) \sum_{j=1}^{n} \frac{\Delta P_j}{\lambda_j - \lambda} \prod_{j+1\le k\le n+1}^{\rightarrow} \left[I + \frac{V\Delta P_k}{\lambda - \lambda_k} \right] d\lambda.$$

Here one can apply the residue theorem.

Furthermore, the following result is proved in [31, Theorem 2.9.1].

Theorem 2.2.3. *For any $A \in \mathbb{C}^{n\times n}$ we have*

$$\lambda R_\lambda = - \prod_{1\le k\le n}^{\rightarrow} \left(I + \frac{A\Delta P_k}{\lambda - \lambda_k} \right) \quad (\lambda \notin \sigma(A) \cup 0), \tag{2.8}$$

where P_k, $k = 1, \dots, n$ is the maximal chain of the invariant projections of A.

Let A be a normal matrix. Then

$$A = \sum_{k=1}^{n} \lambda_k \Delta P_k.$$

Hence, $A\Delta P_k = \lambda_k \Delta P_k$. Since $\Delta P_k \Delta P_j = 0$ for $j \neq k$, the previous theorem gives us the equality

$$-\lambda R_\lambda(A) = I + \sum_{k=1}^{n} \frac{\lambda_k \Delta P_k}{\lambda - \lambda_k} = \sum_{k=1}^{n} \left(\Delta P_k + \frac{\lambda_k \Delta P_k}{\lambda - \lambda_k} \right)$$

or

$$R_\lambda(A) = \sum_{k=1}^{n} \frac{\Delta P_k}{\lambda_k - \lambda}.$$

So from (2.8) we have obtained the well-known spectral representation for the resolvent of a normal matrix. Thus, the previous theorem generalizes the spectral representation for the resolvent of a normal matrix. Now we can use (2.1) and (2.8) to get the representation for $f(A)$.

2.3 Norm estimates for resolvents

For a natural $n \geq 2$, introduce the numbers

$$\gamma_{n,k} = \sqrt{\frac{(n-1)(n-2)\dots(n-k)}{(n-1)^k k!}} \quad (k = 1, 2, \dots, n-1), \gamma_{n,0} = 1.$$

Evidently, for all $n > 2$,

$$\gamma_{n,k}^2 \leq \frac{1}{k!} \quad (k = 1, 2, \dots, n-1). \tag{3.1}$$

Theorem 2.3.1. *Let A be a linear operator in $\mathbb{C}^n$. Then its resolvent satisfies the inequality*

$$\|R_\lambda(A)\| \leq \sum_{k=0}^{n-1} \frac{g^k(A)\gamma_{n,k}}{\rho^{k+1}(A, \lambda)} \quad (\lambda \notin \sigma(A)),$$

where $\rho(A, \lambda) = \min_{k=1,\dots,n} |\lambda - \lambda_k(A)|$.

To prove this theorem again use the Schur triangular representation (2.5). Recall $g(A)$ is defined in Section 2.1. As it is shown in [31, Section 2.1]), the relation $g(U^{-1}AU) = g(A)$ is true, if U is a unitary matrix. Hence it follows that $g(A) = N_2(V)$. The proof of the previous theorem is based on the following lemma.

Lemma 2.3.2. *The inequality*

$$\|R_\lambda(A) - R_\lambda(D)\| \leq \sum_{k=1}^{n-1} \frac{\gamma_{n,k} g^k(A)}{\rho^{k+1}(A,\lambda)} \quad (\lambda \notin \sigma(A))$$

is true.

Proof. By (2.5) we have

$$R_\lambda(A) - R_\lambda(D) = -R_\lambda(D) V R_\lambda(A) = -R_\lambda(D) V (D + V - I\lambda)^{-1}.$$

Thus

$$R_\lambda(A) - R_\lambda(D) = -R_\lambda(D) V (I + R_\lambda(D) V)^{-1} R_\lambda(D). \tag{3.2}$$

Clearly, $R_\lambda(D)V$ is a nilpotent matrix. Hence,

$$R_\lambda(A) - R_\lambda(D) = -R_\lambda(D) V \sum_{k=0}^{n-1} (-R_\lambda(D)V)^k R_\lambda(D) = \sum_{k=1}^{n-1} (-R_\lambda(D)V)^k R_\lambda(D). \tag{3.3}$$

Thanks to Theorem 2.5.1 [31], for any $n \times n$ nilpotent matrix V_0,

$$\|V_0^k\| \leq \gamma_{n,k} N_2^k(V_0). \tag{3.4}$$

In addition, $\|R_\lambda(D)\| = \rho^{-1}(D,\lambda) = \rho^{-1}(A,\lambda)$. So

$$\|(-R_\lambda(D)V)^k\| \leq \gamma_{n,k} N_2^k(R_\lambda(D)V) \leq \frac{\gamma_{n,k} N_2^k(V)}{\rho^k(A,\lambda)}.$$

Now the required result follows from (3.3). □

The assertion of Theorem 2.3.1 directly follows from the previous lemma.

Note that the just proved Lemma 2.3.2 is a slight improvement of Theorem 2.1.1 from [31].

Theorem 2.3.1 is sharp: if A is a normal matrix, then $g(A) = 0$ and Theorem 2.3.1 gives us the equality $\|R_\lambda(A)\| = 1/\rho(A,\lambda)$. Taking into account (3.1), we get

Corollary 2.3.3. *Let $A \in \mathbb{C}^{n \times n}$. Then*

$$\|R_\lambda(A)\| \leq \sum_{k=0}^{n-1} \frac{g^k(A)}{\sqrt{k!}\,\rho^{k+1}(A,\lambda)}$$

for any regular λ of A.

We will need the following result.

Theorem 2.3.4. *Let $A \in \mathbb{C}^{n\times n}$. Then*

$$\|R_\lambda(A)\ \det\,(\lambda I - A)\| \le \left[\frac{N_2^2(A) - 2\,\mathrm{Re}(\overline{\lambda}\ \mathrm{Trace}(A)) \ + n|\lambda|^2}{n-1}\right]^{(n-1)/2}$$

$$(\lambda \notin \sigma(A)).$$

In particular, let V be a nilpotent matrix. Then

$$\|(I\lambda - V)^{-1}\| \le \frac{1}{|\lambda|}\left[1 + \frac{1}{n-1}\left(1 + \frac{N_2^2(V)}{|\lambda|^2}\right)\right]^{(n-1)/2} \qquad (\lambda \ne 0).$$

The proof of this theorem can be found in [31, Section 2.11].

We also point to the following result.

Theorem 2.3.5. *Let $A \in \mathbb{C}^{n\times n}$. Then*

$$\|R_\lambda(A)\| \le \frac{1}{\rho(A,\lambda)}\left[1 + \frac{1}{n-1}\left(1 + \frac{g^2(A)}{\rho^2(A,\lambda)}\right)\right]^{(n-1)/2}$$

for any regular λ of A.

For the proof see [31, Theorem 2.14.1].

2.4 Spectrum perturbations

Let A and B be $n \times n$-matrices having eigenvalues

$$\lambda_1(A), \dots, \lambda_n(A) \quad \text{and} \quad \lambda_1(B), \dots, \lambda_n(B),$$

respectively, and $q = \|A - B\|$.

The spectral variation of B with respect to A is

$$sv_A(B) := \max_i \min_j |\lambda_i(B) - \lambda_j(A)|,$$

cf. [103]. The following simple lemma is proved in [31, Section 4.1].

Lemma 2.4.1. *Assume that $\|R_\lambda(A)\| \le \phi(\rho^{-1}(A,\lambda))$ for all regular λ of A, where $\phi(x)$ is a monotonically increasing non-negative continuous function of a non-negative variable x, such that $\phi(0) = 0$ and $\phi(\infty) = \infty$. Then the inequality $sv_A(B) \le z(\phi, q)$ is true, where $z(\phi, q)$ is the unique positive root of the equation $q\phi(1/z) = 1$.*

This lemma and Corollary 2.3.2 yield our next result.

Theorem 2.4.2. *Let A and B be $n \times n$-matrices. Then $sv_A(B) \le z(q,A)$, where $z(q,A)$ is the unique non-negative root of the algebraic equation*

$$y^n = q \sum_{j=0}^{n-1} \frac{y^{n-j-1} g^j(A)}{\sqrt{j!}}. \tag{4.1}$$

Let us consider the algebraic equation

$$z^n = P(z) \quad (n > 1), \quad \text{where} \quad P(z) = \sum_{j=0}^{n-1} c_j z^{n-j-1} \tag{4.2}$$

with non-negative coefficients c_j $(j = 0, \dots, n-1)$.

Lemma 2.4.3. *The extreme right-hand root z_0 of equation (4.2) is non-negative and the following estimates are valid:*

$$z_0 \le \sqrt[n]{P(1)} \ \textit{if} \ P(1) \le 1, \tag{4.3}$$

and

$$1 \le z_0 \le P(1) \ \textit{if} \ P(1) \ge 1. \tag{4.4}$$

Proof. Since all the coefficients of $P(z)$ are non-negative, it does not decrease as $z > 0$ increases. From this it follows that if $P(1) \le 1$, then $z_0 \le 1$. So $z_0^n \le P(1)$, as claimed.

Now let $P(1) \ge 1$, then due to (4.2) $z_0 \ge 1$ because $P(z)$ does not decrease. It is clear that

$$P(z_0) \le z_0^{n-1} P(1)$$

in this case. Substituting this inequality into (4.2), we get (4.4). □

Substituting $z = ax$ with a positive constant a into (4.2), we obtain

$$x^n = \sum_{j=0}^{n-1} \frac{c_j}{a^{j+1}} x^{n-j-1}. \tag{4.5}$$

Let

$$a = 2 \max_{j=0,\dots,n-1} \sqrt[j+1]{c_j}.$$

Then

$$\sum_{j=0}^{n-1} \frac{c_j}{a^{j+1}} \le \sum_{j=0}^{n-1} 2^{-j-1} = 1 - 2^{-n} < 1.$$

Let x_0 be the extreme right-hand root of equation (4.5), then by (4.3) we have $x_0 \le 1$. Since $z_0 = ax_0$, we have derived the following result.

Corollary 2.4.4. *The extreme right-hand root z_0 of equation* (4.2) *is non-negative. Moreover,*

$$z_0 \le 2 \max_{j=0,\dots,n-1} \sqrt[j+1]{c_j}.$$

Now put $y = xg(A)$ into (6.1). Then we obtain the equation

$$x^n = \frac{q}{g(A)} \sum_{j=0}^{n-1} \frac{x^{n-j-1}}{\sqrt{j!}}.$$

Put

$$w_n = \sum_{j=0}^{n-1} \frac{1}{\sqrt{j!}}.$$

Applying Lemma 2.4.3 we get the estimate $z(q, A) \le \delta(q)$, where

$$\delta(q) := \begin{cases} qw_n & \text{if } qw_n \ge g(A), \\ g^{1-1/n}(A)[qw_n]^{1/n} & \text{if } qw_n \le g(A). \end{cases}$$

Now Theorem 2.4.2 ensures the following result.

Corollary 2.4.5. *One has $sv_A(B) \le \delta(q)$.*

Furthermore let $\tilde{D}, V_+$ and V_- be the diagonal, upper nilpotent part and lower nilpotent part of matrix A, respectively. Using the notation $A_+ = \tilde{D} + V_+$, we arrive at the relations

$$\sigma(A_+) = \sigma(\tilde{D}), g(A_+) = N_2(V_+) \quad \text{and} \quad \|A - A_+\| = \|V_-\|.$$

Taking

$$\delta_A := \begin{cases} \|V_-\|w_n & \text{if } \|V_-\|w_n \ge N_2(V_+), \\ N_2^{1-1/n}(V_+)[\|V_-\|w_n]^{1/n} & \text{if } \|V_-\|w_n \le N_2(V_+), \end{cases}$$

due to the previous corollary we obtain

Corollary 2.4.6. *Let $A = (a_{jk})_{j,k=1}^n$ be an $n \times n$-matrix. Then for any eigenvalue μ of A, there is a $k = 1, \dots, n$, such that*

$$|\mu - a_{kk}| \le \delta_A,$$

and therefore the (upper) spectral radius satisfies the inequality

$$r_s(A) \le \max_{k=1,\dots,n} |a_{kk}| + \delta_A,$$

and the lower spectral radius satisfies the inequality

$$r_l(A) \ge \min_{k=1,\dots,n} |a_{kk}| - \delta_A,$$

provided $|a_{kk}| > \delta_A$ $(k = 1, \dots, n)$.

Clearly, one can exchange the places V_+ and V_-.

Let us recall the celebrated Gerschgorin theorem. To this end write

$$R_j = \sum_{k=1,k\neq j}^{n} |a_{jk}|.$$

Let $\Omega(b,r)$ be the closed disc centered at $b \in \mathbb{C}$ with a radius r.

Theorem 2.4.7 (Gerschgorin). *Every eigenvalue of A lies within at least one of the discs $\Omega(a_{jj}, R_j)$.*

Proof. Let λ be an eigenvalue of A and let $x = (x_j)$ be the corresponding eigenvector. Let i be chosen so that $|x_i| = \max_j |x_j|$. Then $|x_i| > 0$, otherwise $x = 0$. Since x is an eigenvector, $Ax = \lambda x$ or equivalent

$$\sum_{k=1}^{n} a_{ik} = \lambda x_i$$

so, splitting the sum, we get

$$\sum_{k=1,k\neq i}^{n} a_{ik} x_k = \lambda x_i - a_{ii} x_i.$$

We may then divide both sides by x_i (choosing i as we explained we can be sure that $x_i \neq 0$) and take the absolute value to obtain

$$|\lambda - a_{ii}| \leq \sum_{k=1,k\neq i}^{n} |a_{jk}| \frac{|x_k|}{|x_i|} \leq R_i,$$

where the last inequality is valid because

$$\frac{|x_k|}{|x_l|} \leq 1$$

as claimed. □

Note that for a diagonal matrix the Gerschgorin discs $\Omega(a_{jj}, R_j)$ coincide with the spectrum. Conversely, if the Gerschgorin discs coincide with the spectrum, the matrix is diagonal.

The next lemma follows from the Gerschgorin theorem and gives us a simple bound for the spectral radius.

Lemma 2.4.8. *The spectral radius $r_s(A)$ of a matrix $A = (a_{jk})_{j,k=1}^n$ satisfies the inequality*

$$r_s(A) \leq \max_j \sum_{k=1}^{n} |a_{jk}|.$$

About this and other estimates for the spectral radius see [80, Section 16].

2.5 Norm estimates for matrix functions

2.5.1 Estimates via the resolvent

The following result directly follows from (2.1).

Lemma 2.5.1. *Let let $f(\lambda)$ be a scalar-valued function which is regular on a neighborhood M of an open simply-connected set containing the spectrum of $A \in \mathbb{C}^{n\times n}$, and $C \subset M$ be a closed smooth contour surrounding $\sigma(A)$. Then*

$$\|f(A)\| \leq \frac{1}{2\pi}\int_C |f(z)|\|R_z(A)\|dz \leq m_C(A) l_C \sup_{z\in C}|f(z)|,$$

where

$$m_C(A) := \sup_{z\in C}\|R_z(A)\|,\ l_C := \frac{1}{2\pi}\int_C |dz|.$$

Now we can directly apply the estimates for the resolvent from Section 2.3. In particular, by Corollary 2.3.3 we have

$$\|R_z(A)\| \leq p(A, 1/\rho(A,z)), \tag{5.1}$$

where

$$p(A,x) = \sum_{k=0}^{n-1} \frac{x^{k+1} g^k(A)}{\sqrt{k!}} \quad (x>0). \tag{5.2}$$

We thus get $m_C(A) \leq p(A, 1/\rho(A,C))$, where $\rho(A,C)$ is the distance between C and $\sigma(A)$, and therefore,

$$\|f(A)\| \leq l_C p(A, 1/\rho(A,C)) \sup_{z\in C}|f(z)|. \tag{5.3}$$

2.5.2 Functions regular on the convex hull of the spectrum

Theorem 2.5.2. *Let A be an $n\times n$-matrix and f be a function holomorphic on a neighborhood of the convex hull* $\mathrm{co}(A)$ *of $\sigma(A)$. Then*

$$\|f(A)\| \leq \sup_{\lambda\in\sigma(A)}|f(\lambda)| + \sum_{k=1}^{n-1} \sup_{\lambda\in \mathrm{co}(A)} |f^{(k)}(\lambda)| \frac{\gamma_{n,k} g^k(A)}{k!}.$$

This theorem is proved in the next subsection. Taking into account (3.1) we get our next result.

Corollary 2.5.3. *Under the hypothesis of Theorem* 2.5.2 *we have*

$$\|f(A)\| \leq \sup_{\lambda\in\sigma(A)}|f(\lambda)| + \sum_{k=1}^{n-1} \sup_{\lambda\in \mathrm{co}(A)} |f^{(k)}(\lambda)| \frac{g^k(A)}{(k!)^{3/2}}.$$

Theorem 2.5.2 is sharp: if A is normal, then $g(A) = 0$ and

$$\|f(A)\| = \sup_{\lambda \in \sigma(A)} |f(\lambda)|.$$

For example,

$$\|\exp(At)\| \leq e^{\alpha(A)t} \sum_{k=0}^{n-1} g^k(A) t^k \frac{\gamma_{n,k}}{k!} \leq e^{\alpha(A)t} \sum_{k=0}^{n-1} \frac{g^k(A)t^k}{(k!)^{3/2}} \quad (t \geq 0)$$

where $\alpha(A) = \max_{k=1,\dots,n} \operatorname{Re} \lambda_k(A)$. In addition,

$$\|A^m\| \leq \sum_{k=0}^{n-1} \frac{\gamma_{n,k} m! g^k(A) r_s^{m-k}(A)}{(m-k)!k!} \leq \sum_{k=0}^{n-1} \frac{m! g^k(A) r_s^{m-k}(A)}{(m-k)!(k!)^{3/2}} \quad (m = 1, 2, \dots),$$

where $r_s(A)$ is the spectral radius. Recall that $1/(m-k)! = 0$ if $m < k$.

2.5.3 Proof of Theorem 2.5.2

Let $|V|_e$ be the operator whose entries in the orthonormal basis of the triangular representation (the Schur basis) $\{e_k\}$ are the absolute values of the entries of the nilpotent part V of A with respect to this basis. That is,

$$|V|_e = \sum_{k=1}^{n} \sum_{j=1}^{k-1} |a_{jk}| (., e_k) e_j,$$

where $a_{jk} = (Ae_k, e_j)$. Put

$$I_{j_1 \dots j_{k+1}} = \frac{(-1)^{k+1}}{2\pi i} \int_C \frac{f(\lambda) d\lambda}{(\lambda_{j_1} - \lambda) \dots (\lambda_{j_{k+1}} - \lambda)}.$$

We need the following result.

Lemma 2.5.4. *Let A be an $n \times n$-matrix and f be a holomorphic function on a Jordan domain (that is on a closed simply connected set, whose boundary is a Jordan contour), containing $\sigma(A)$. Let D be the diagonal part of A. Then*

$$\|f(A) - f(D)\| \leq \sum_{k=1}^{n-1} J_k \| \, |V|_e^k \|,$$

where

$$J_k = \max\{|I_{j_1 \dots j_{k+1}}| : 1 \leq j_1 < \dots < j_{k+1} \leq n\}.$$

Proof. From (2.1) and (3.3) we deduce that

$$f(A) - f(D) = -\frac{1}{2\pi i}\int_C f(\lambda)(R_\lambda(A) - R_\lambda(D))d\lambda = \sum_{k=1}^{n-1} B_k, \tag{5.4}$$

where

$$B_k = (-1)^{k+1}\frac{1}{2\pi i}\int_C f(\lambda)(R_\lambda(D)V)^k R_\lambda(D)d\lambda.$$

Since D is a diagonal matrix with respect to the Schur basis $\{e_k\}$ and its diagonal entries are the eigenvalues of A, then

$$R_\lambda(D) = \sum_{j=1}^{n} \frac{\Delta P_j}{\lambda_j(A) - \lambda},$$

where $\Delta P_k = (., e_k)e_k$. In addition, $\Delta P_j V \Delta P_k = 0$ for $j \geq k$. Consequently,

$$B_k = \sum_{j_1=1}^{j_2-1} \Delta P_{j_1} V \sum_{j_2=1}^{j_3-1} \Delta P_{j_2} V \cdots \sum_{j_k=1}^{j_{k+1}-1} V \sum_{j_{k+1}=1}^{n} \Delta P_{j_{k+1}} I_{j_1 j_2 \dots j_{k+1}}.$$

Lemma 2.8.1 from [31] gives us the estimate

$$\begin{aligned} \|B_k\| &\leq J_k \| \sum_{j_1=1}^{j_2-1} \Delta P_{j_1} |V|_e \sum_{j_2=1}^{j_3-1} \Delta P_{j_2} |V|_e \cdots \sum_{j_k=1}^{j_{k+1}-1} |V|_e \sum_{j_{k+1}=1}^{n} \Delta P_{j_{k+1}} \| \\ &= J_k \| P_{n-k}|V|_e P_{n-k+1}|V|_e P_{n-k+2} \dots P_{n-1}|V|_e \|. \end{aligned}$$

But

$$\begin{aligned} P_{n-k}|V|_e P_{n-k+1}|V|_e P_{n-k+2} \dots P_{n-1}|V|_e &= |V|_e P_{n-k+1}|V|_e P_{n-k+2} \dots P_{n-1}|V|_e \\ &= |V|_e^2 \dots P_{n-1}|V|_e = |V|_e^k. \end{aligned}$$

Thus

$$\|B_k\| \leq J_k \| \, |V|_e^k \, \|.$$

This inequality and (5.4) imply the required inequality. □

Since $N_2(|V|_e) = N_2(V) = g(A)$, by (3.4) and the previous lemma we get the following result.

Lemma 2.5.5. *Under the hypothesis of Lemma* 5.4 *we have*

$$\|f(A) - f(D)\| \leq \sum_{k=1}^{n-1} J_k \gamma_{n,k} g^k(A).$$

Let f be holomorphic on a neighborhood of co(A). Thanks to Lemma 1.5.1 from [31],

$$J_k \leq \frac{1}{k!} \sup_{\lambda \in \mathrm{co}(A)} |f^{(k)}(\lambda)|.$$

Now the previous lemma implies

Corollary 2.5.6. *Under the hypothesis of Theorem* 2.5.2 *we have the inequalities*

$$\|f(A) - f(D)\| \leq \sum_{k=1}^{n-1} \sup_{\lambda \in \mathrm{co}(A)} |f^{(k)}(\lambda)| \gamma_{n,k} \frac{g^k(A)}{k!}.$$

The assertion of Theorem 2.5.2 directly follows from the previous corollary.

Note that the latter corollary is a slight improvement of Theorem 2.7.1 from [31].

Denote by $f[a_1, a_2, \ldots, a_{k+1}]$ the kth divided difference of f at points a_1, a_2, $\ldots$, a_{k+1}. By the Hadamard representation [20, formula (54)], we have

$$I_{j_1 \ldots j_{k+1}} = f[\lambda_1, \ldots, \lambda_{j_{k+1}}],$$

provided λ_j are distinct. Now Lemma 5.5 implies

Corollary 2.5.7. *Let all the eigenvalues of an $n \times n$-matrix A be algebraically simple, and f be a holomorphic function in a Jordan domain containing $\sigma(A)$. Then*

$$\|f(A) - f(D)\| \leq \sum_{k=1}^{n-1} f_k \gamma_{n,k} g^k(A) \leq \sum_{k=1}^{n-1} f_k \frac{g^k(A)}{\sqrt{k!}},$$

where

$$f_k = \max\{|f[\lambda_1(A), \ldots, \lambda_{j_{k+1}}(A)]| : 1 < j_1 < \cdots < j_{k+1} \leq n\}.$$

2.6 Absolute values of entries of matrix functions

Everywhere in the present section, $A = (a_{jk})_{j,k=1}^n$, $S = \mathrm{diag}[a_{11}, \ldots, a_{nn}]$ and the off diagonal of A is $W = A - S$. That is, the entries v_{jk} of W are $v_{jk} = a_{jk}$ $(j \neq k)$ and $v_{jj} = 0$ $(j, k = 1, 2, \ldots)$. Denote by co(S) the closed convex hull of the diagonal entries $a_{11}, \ldots, a_{nn}$. We put $|A| = (|a_{jk}|)_{j,l=1}^n$, i.e., $|A|$ is the matrix whose entries are the absolute values of the entries A in the standard basis. We also write $T \geq 0$ if all the entries of a matrix T are non-negative. If T and B are two matrices, then we write $T \geq B$ if $T - B \geq 0$.

Thanks to Lemma 2.4.8 we obtain $r_s(|W|) \leq \tau_W$, where

$$\tau_W := \max_j \sum_{k=1, k \neq j}^{n} |a_{jk}|.$$

Theorem 2.6.1. *Let $f(\lambda)$ be holomorphic on a neighborhood of a Jordan set, whose boundary C has the property*

$$|z - a_{jj}| > \sum_{k=1,k\neq j}^{n} |a_{jk}| \tag{6.1}$$

for all $z \in C$ and $j = 1, \ldots, n$. Then, with the notation

$$\xi_k(A) := \sup_{z \in \mathrm{co}\,(S)} \frac{|f^{(k)}(z)|}{k!} \quad (k = 1, 2, \ldots),$$

the inequality

$$|f(A) - f(S)| \leq \sum_{k=1}^{\infty} \xi_k(A) |W|^k$$

is valid, provided

$$r_s(|W|) \lim_{k \to \infty} \sqrt[k]{\xi_k(A)} < 1.$$

Proof. By the equality $A = S + W$ we get

$$R_\lambda(A) = (S + W - \lambda I)^{-1} = (I + R_\lambda(S)W)^{-1} R_\lambda(S) = \sum_{k=0}^{\infty} (R_\lambda(S)W)^k (-1)^k R_\lambda(S),$$

provided the spectral radius $r_0(\lambda)$ of $R_\lambda(S)W$ is less than one. The entries of this matrix are

$$\frac{a_{jk}}{a_{jj} - \lambda} \quad (\lambda \neq a_{jj}, \ j \neq k)$$

and the diagonal entries are zero. Thanks to Lemma 2.4.8 we have

$$r_0(\lambda) \leq \max_j \sum_{k=1,k\neq j}^{n} \frac{|a_{jk}|}{|a_{jj} - \lambda|} < 1 \ \ (\lambda \in C)$$

and the series

$$R_\lambda(A) - R_\lambda(S) = \sum_{k=1}^{\infty} (R_\lambda(S)W)^k (-1)^k R_\lambda(S)$$

converges. Thus

$$f(A) - f(S) = -\frac{1}{2\pi i} \int_C f(\lambda)(R_\lambda(A) - R_\lambda(S)) d\lambda = \sum_{k=1}^{\infty} M_k, \tag{6.2}$$

where

$$M_k = (-1)^{k+1} \frac{1}{2\pi i} \int_C f(\lambda)(R_\lambda(S)W)^k R_\lambda(S) d\lambda.$$

Since S is a diagonal matrix with respect to the standard basis $\{d_k\}$, we can write

$$R_\lambda(S) = \sum_{j=1}^{n} \frac{\hat{Q}_j}{b_j - \lambda} \quad (b_j = a_{jj}),$$

where $\hat{Q}_k = (., d_k)d_k$. We thus have

$$M_k = \sum_{j_1=1}^{n} \hat{Q}_{j_1} W \sum_{j_2=1}^{n} \hat{Q}_{j_2} W \dots W \sum_{j_{k+1}=1}^{n} \hat{Q}_{j_{k+1}} J_{j_1 j_2 \dots j_{k+1}}. \tag{6.3}$$

Here

$$J_{j_1 \dots j_{k+1}} = \frac{(-1)^{k+1}}{2\pi i} \int_C \frac{f(\lambda) d\lambda}{(b_{j_1} - \lambda) \dots (b_{j_{k+1}} - \lambda)}.$$

Lemma 1.5.1 from [31] gives us the inequalities

$$|J_{j_1 \dots j_{k+1}}| \le \xi_k(A) \ \ (j_1, j_2, \dots, j_{k+1} = 1, \dots, n).$$

Hence, by (6.3)

$$|M_k| \le \xi_k(A) \sum_{j_1=1}^{n} \hat{Q}_{j_1} |W| \sum_{j_2=1}^{n} \hat{Q}_{j_2} |W| \dots |W| \sum_{j_{k+1}=1}^{n} \hat{Q}_{j_{k+1}}.$$

But

$$\sum_{j_1=1}^{n} \hat{Q}_{j_1} |W| \sum_{j_2=1}^{n} \hat{Q}_{j_2} |W| \dots |W| \sum_{j_{k+1}=1}^{n} \hat{Q}_{j_{k+1}} = |W|^k.$$

Thus $|M_k| \le \xi_k(A)|W|^k$. Now (6.2) implies the required result. □

Additional estimates for the entries of matrix functions can be found in [43, 38]. Under the hypothesis of the previous theorem with the notation

$$\xi_0(A) := \max_k |f(a_{kk})|,$$

we have the inequality

$$|f(A)| \le \xi_0(A) I + \sum_{k=1}^{\infty} \xi_k(A) |W|^k = \sum_{k=0}^{\infty} \xi_k(A) |W|^k. \tag{6.4}$$

Here $|W|^0 = I$.

Let $\|A\|_l$ denote a lattice norm of A. That is, $\|A\|_l \le \||A|\|_l$, and $\|A\|_l \le \|\tilde{A}\|_l$ whenever $0 \le A \le \tilde{A}$. Now the previous theorem implies the inequality

$$\|f(A) - f(S)\|_l \le \sum_{k=1}^{\infty} \xi_k(A) \||W|^k\|_l \tag{6.5}$$

and therefore,

$$\|f(A)\|_l \le \sum_{k=0}^{\infty} \xi_k(A) \||W|^k\|_l.$$

2.7 Diagonalizable matrices

The results of this section are useful for investigation of the forced oscillations of equations close to ordinary ordinary differential equations (see Section 12.3).

2.7.1 A bound for similarity constants of matrices

Everywhere in this section it is assumed that the eigenvalues $\lambda_k = \lambda_k(A)$ $(k = 1, \ldots, n)$ of A, taken with their algebraic multiplicities, are geometrically simple. That is, the geometric multiplicity of each eigenvalue is equal to one. As it is well known, in this case A is diagonalizable: there are biorthogonal sequences $\{u_k\}$ and $\{v_k\}$: $(v_j, u_k) = 0$ $(j \neq k)$, $(v_j, u_j) = 1$ $(j, k = 1, \ldots, n)$, such that

$$A = \sum_{k=1}^{n} \lambda_k Q_k, \tag{7.1}$$

where $Q_k = (., u_k) v_k$ $(k = 1, \ldots, n)$ are one-dimensional eigenprojections. Besides, there is an invertible operator T and a normal operator S, such that

$$TA = ST. \tag{7.2}$$

The constant (the conditional number)

$$\kappa_T := \|T\| \, \|T^{-1}\|$$

is very important for various applications, cf. [103]. That constant is mainly numerically calculated. In the present subsection we suggest a sharp bound for κ_T. Applications of the obtained bound are also discussed.

Denote by $\mu_j, j = 1, \ldots, m \leq n$ the distinct eigenvalues of A, and by p_j the algebraic multiplicity of μ_j. In particular, one can write

$$\mu_1 = \lambda_1 = \cdots = \lambda_{p_1}, \; \mu_2 = \lambda_{p_1+1} = \cdots = \lambda_{p_1+p_2},$$

etc.

Let δ_j be the half-distance from μ_j to the other eigenvalues of A:

$$\delta_j := \min_{k=1,\ldots,m;\; k \neq j} |\mu_j - \mu_k|/2$$

and

$$\delta(A) := \min_{j=1,\ldots,m} \delta_j = \min_{j,k=1,\ldots,m;\; k \neq j} |\mu_j - \mu_k|/2.$$

Put

$$\eta\,(A) := \sum_{k=1}^{n-1} \frac{g^k(A)}{\delta^k(A)\sqrt{k!}}.$$

According to (1.1),

$$\eta(A) \leq \sum_{k=1}^{n-1} \frac{(\sqrt{2}N_2(A_I))^k}{\delta^k(A)\sqrt{k!}}.$$

In [51, Corollary 3.6], the inequality

$$\kappa_T \leq \sum_{j=1}^{m} p_j \sum_{k=0}^{n-1} \frac{g^k(A)}{\delta_j^k \sqrt{k!}} \leq n(1+\eta(A)) \tag{7.3}$$

has been derived. This inequality is not sharp: if A is a normal matrix, then it gives $\kappa_T \leq n$ but $\kappa_T = 1$ in this case. In this section we improve inequality (7.3). To this end put

$$\gamma(A) = \begin{cases} n(1+\eta(A)) & \text{if } \eta(A) \geq 1, \\ (\eta(A)+1)[\frac{2\sqrt{n}\eta(A)}{1-\eta(A)}+1] & \text{if } \eta(A) < 1. \end{cases}$$

Now we are in a position to formulate the main result of the present section.

Theorem 2.7.1. *Let A be a diagonalizable $n \times n$-matrix. Then $\kappa_T \leq \gamma(A)$.*

The proof of this theorem is presented in the next subsection. Theorem 2.7.1 is sharp: if A is normal, then $g(A) = 0$. Therefore $\eta(A) = 0$ and $\gamma(A) = 1$. Thus we obtain the equality $\kappa_T = 1$.

2.7.2 Proof of Theorem 2.7.1

We need the following lemma [51, Lemma 3.4].

Lemma 2.7.2. *Let A be a diagonalizable $n \times n$-matrix and*

$$S = \sum_{k=1}^{n} \lambda_k(., d_k)d_k, \tag{7.4}$$

where $\{d_k\}$ is an orthonormal basis. Then the operator

$$T = \sum_{k=1}^{n} (., d_k)v_k \tag{7.5}$$

has the inverse one defined by

$$T^{-1} = \sum_{k=1}^{n} (., u_k)d_k,$$

and (7.2) holds.

Note that one can take $\|u_k\| = \|v_k\|$. This leads to the equality

$$\|T^{-1}\| = \|T\|. \tag{7.6}$$

We need also the following technical lemma.

Lemma 2.7.3. *Let L_1 and L_2 be projections satisfying the condition $r := \|L_1 - L_2\| < 1$. Then for any eigenvector f_1 of L_1 with $\|f_1\| = 1$ and $L_1 f_1 = f_1$, there exists an eigenvector f_2 of L_2 with $\|f_2\| = 1$ and $L_2 f_2 = f_2$, such that*

$$\|f_1 - f_2\| \le \frac{2r}{1-r}.$$

Proof. We have $\|L_2 f_1 - L_1 f_1\| \le r < 1$ and

$$b_0 := \|L_2 f_1\| \ge \|L_1 f_1\| - \|(L_1 - L_2) f_1\| \ge 1 - r > 0.$$

Thanks to the relation $L_2(L_2 f_1) = L_2 f_1$, we can assert that $L_2 f_1$ is an eigenvector of L_2. Then

$$f_2 := \frac{1}{b_0} L_2 f_1$$

is a normed eigenvector of L_2. So

$$f_1 - f_2 = L_1 f_1 - \frac{1}{b_0} L_2 f_1 = f_1 - \frac{1}{b_0} f_1 + \frac{1}{b_0}(L_1 - L_2) f_1.$$

But

$$\frac{1}{b_0} \le \frac{1}{1-r}$$

and

$$\|f_1 - f_2\| \le \left(\frac{1}{b_0} - 1\right)\|f_1\| + \frac{1}{b_0}\|(L_1 - L_2) f_1\| \le \frac{1}{1-r} - 1 + \frac{r}{1-r} = \frac{2r}{1-r},$$

as claimed. □

In the rest of this subsection $d_k = e_k$ $(k = 1, \dots n)$, where $\{e_k\}_{k=1}^n$ is the Schur orthonormal basis of A. Then $S = D$, where D is the diagonal part of A (see Section 2.2). For a fixed index $j \le m$, let $\hat{P}_j$ be the eigenprojection of D, and $\hat{Q}_j$ the eigenprojection of A corresponding to the same (geometrically simple) eigenvalue μ_j. So

$$A = \sum_{j=1}^{m} \mu_j \hat{Q}_j.$$

The following inequality is proved in [51, inequality (3.3)]:

$$\|\hat{Q}_j - \hat{P}_j\| \le \eta_j, \quad \text{where} \quad \eta_j := \sum_{k=1}^{n-1} \frac{g^k(A)}{\delta_j^k \sqrt{k!}}. \tag{7.7}$$

Let v_{js} and e_{js} $(s = 1, \dots, p_j)$ be the eigenvectors of $\hat{Q}_j$ and $\hat{P}_j$, respectively, and $\|e_{js}\| = 1$. Inequality (7.7) and the previous lemma yield

Corollary 2.7.4. *Assume that*

$$\eta(A) < 1. \tag{7.8}$$

Then

$$\left\| \frac{v_{js}}{\|v_{js}\|} - e_{js} \right\| \le \psi(A) \;\; (s = 1, \dots, p_j), \quad \textit{where} \quad \psi(A) = \frac{2\eta(A)}{1 - \eta(A)}.$$

Hence it follows that

$$\left\| \frac{v_k}{\|v_k\|} - e_k \right\| \le \psi(A) \;\; (k = 1, \dots, n).$$

Taking in the previous corollary Q_j^* instead of Q_j we arrive at the similar inequality

$$\left\| \frac{u_k}{\|u_k\|} - e_k \right\| \le \psi(A) \;\; (k = 1, \dots, n). \tag{7.9}$$

Proof of Theorem 2.7.1. By (7.5) we have

$$\|Tx\| = \left[\sum_{k=1}^{n} |(x, u_k)|^2 \right]^{1/2} = \left[\sum_{k=1}^{n} |(x, \|u_k\| e_k) + (x, u_k - \|u_k\| e_k)|^2 \right]^{1/2}$$
$$\le \left[\sum_{k=1}^{n} \|u_k\|^2 |(x, \frac{u_k}{\|u_k\|} - e_k)|^2 \right]^{1/2} + \left[\sum_{k=1}^{n} |\|u_k\| (x, e_k)|^2 \right]^{1/2} \quad (x \in \mathbb{C}^n).$$

Hence, under condition (7.8), inequality (7.9) implies

$$\|Tx\| \le \max_k \|u_k\| \|x\| (\psi(A)\sqrt{n} + 1) \;\; (x \in \mathbb{C}^n).$$

Thanks to Corollary 3.3 from [51], $\max_k \|u_k\|^2 \le 1 + \eta(A)$. Thus,

$$\|T\|^2 \le \tilde{\gamma}(A) := (\eta(A) + 1) \left[\frac{2\sqrt{n}\eta(A)}{1 - \eta(A)} + 1 \right].$$

According to (7.6), $\|T^{-1}\|^2 \le \tilde{\gamma}(A)$. So under condition (7.8), the inequality $\kappa_T \le \tilde{\gamma}(A)$ is true. Combining this inequality with (7.3), we get the required result. □

2.7.3 Applications of Theorem 2.7.1

Theorem 2.7.1 immediately implies

Corollary 2.7.5. *Let A be a diagonalizable $n \times n$-matrix and $f(z)$ be a scalar function defined on the spectrum of A. Then $\|f(A)\| \le \gamma(A) \max_k |f(\lambda_k)|$.*

In particular, we have

$$\|R_z(A)\| \leq \frac{\gamma(A)}{\rho(A,\lambda)} \quad \text{and} \quad \|e^{At}\| \leq \gamma(A)e^{\alpha(A)t} \ (t \geq 0).$$

Let A and $\tilde{A}$ be complex $n \times n$-matrices whose eigenvalues λ_k and $\tilde{\lambda}_k$, respectively, are taken with their algebraic multiplicities. Recall that

$$sv_A(\tilde{A}) := \max_k \min_j |\tilde{\lambda}_k - \lambda_j|.$$

Corollary 2.7.6. *Let A be diagonalizable. Then $sv_A(\tilde{A}) \leq \gamma(A)\|A - \tilde{A}\|$.*

Indeed, the operator $S = TAT^{-1}$ is normal. Put $B = T\tilde{A}T^{-1}$. Thanks to the well-known Corollary 3.4 [103], $sv_S(B) \leq \|S - B\|$. Now the required result is due to Theorem 2.7.1.

Furthermore let $\tilde{D}, V_+$ and V_- be the diagonal, upper nilpotent part and lower nilpotent part of matrix $A = (a_{kk})$, respectively. Using the preceding corollary with $A_+ = \tilde{D} + V_+$, we arrive at the relations

$$\sigma(A_+) = \sigma(\tilde{D}), \quad \text{and} \quad \|A - A_+\| = \|V_-\|.$$

Due to the previous corollary we get

Corollary 2.7.7. *Let $A = (a_{jk})_{j,k=1}^n$ be an $n \times n$-matrix, whose diagonal has the property*

$$a_{jj} \neq a_{kk} \ (j \neq k; \ k = 1, \dots, n).$$

Then for any eigenvalue μ of A, there is a $k = 1, \dots, n$, such that

$$|\mu - a_{kk}| \leq \gamma(A_+)\|V_-\|,$$

and therefore the (upper) spectral radius satisfies the inequality

$$r_s(A) \leq \max_{k=1,\dots,n} |a_{kk}| + \gamma(A_+)\|V_-\|,$$

and the lower spectral radius satisfies the inequality

$$r_s(A) \geq \min_{k=1,\dots,n} |a_{kk}| - \gamma(A_+)\|V_-\|,$$

provided $|a_{kk}| > \delta_A$ $(k = 1, \dots, n)$.

Clearly, one can exchange the places V_+ and V_-.

2.7.4 Additional norm estimates for functions of diagonalizable matrices

Again A is a diagonalizable $n \times n$-matrix, and μ_j $(j = 1, \dots, m \leq n)$ are the distinct eigenvalues of A: $\mu_j \neq \mu_k$ for $k \neq j$ and enumerated in an arbitrary order. Thus, (7.1) can be written as

$$A = \sum_{k=1}^{m} \mu_k \hat{Q}_k,$$

where $\hat{Q}_k, k \leq m$, is the eigenprojection whose dimension is equal to the algebraic multiplicity of μ_k.

Besides, for any function f defined on $\sigma(A)$, $f(A)$ can be represented as

$$f(A) = \sum_{k=1}^{m} f(\mu_k) \hat{Q}_k.$$

Put

$$\tilde{Q}_j = \sum_{k=1}^{j} \hat{Q}_k \quad (j = 1, \dots, m).$$

Note that according to (7.2) $\tilde{Q}_j = T^{-1} \hat{P}_j T$, where $\hat{P}_j$ is an orthogonal projection. Thus by Theorem 2.7.1,

$$\max_{1 \leq k \leq m} \|\tilde{Q}_k\| \leq \kappa_T \leq \gamma(A).$$

Theorem 2.7.8. *Let A be diagonalizable. Then*

$$\|f(A) - f(\mu_m) I\| \leq \max_{1 \leq k \leq m} \|\tilde{Q}_k\| \sum_{k=1}^{m-1} |f(\mu_k) - f(\mu_{k+1})|.$$

To prove this theorem, we need the following analog of the Abel transform.

Lemma 2.7.9. *Let a_k be numbers and W_k $(k = 1, \dots, m)$ be bounded linear operators in a Banach space. Let*

$$\Psi := \sum_{k=1}^{m} a_k W_k \quad \textit{and} \quad B_j = \sum_{k=1}^{j} W_k.$$

Then

$$\Psi = \sum_{k=1}^{m-1} (a_k - a_{k+1}) B_k + a_m B_m.$$

Proof. Obviously,

$$\Psi = a_1 B_1 + \sum_{k=2}^{m} a_k (B_k - B_{k-1}) = \sum_{k=1}^{m} a_k B_k - \sum_{k=2}^{m} a_k B_{k-1} = \sum_{k=1}^{m} a_k B_k - \sum_{k=1}^{m-1} a_{k+1} B_k,$$

as claimed. □

Proof of Theorem 2.7.8. According to Lemma 2.7.9,

$$f(A) = \sum_{k=1}^{m-1} (f(\mu_k) - f(\mu_{k+1}))\tilde{Q}_k + f(\mu_m)\tilde{Q}_m. \tag{7.10}$$

But $\tilde{Q}_m = I$. Hence, the assertion of the theorem at once follows. □

According to the previous theorem we have

Corollary 2.7.10. *Assume that $f(\lambda)$ is real for all $\lambda \in \sigma(A)$, and that the condition*

$$f(\mu_{k+1}) \leq f(\mu_k) \quad (k = 1, \dots, m-1) \tag{7.11}$$

holds. Then

$$\|f(A) - f(\mu_m)I\| \leq \max_{1 \leq k \leq m} \|\tilde{Q}_k\| [f(\mu_1) - f(\mu_m)].$$

From this corollary it follows that

$$\|f(A)\| \leq \max_{1 \leq k \leq m} \|\tilde{Q}_k\| [f(\mu_1) + |f(\mu_m)| - f(\mu_m)],$$

provided (7.11) holds.

2.8 Matrix exponential

2.8.1 Lower bounds

As it was shown in Section 2.5,

$$\|\exp(At)\| \leq e^{\alpha(A)t} \sum_{k=0}^{n-1} g^k(A) t^k \frac{\gamma_{n,k}}{k!} \quad (t \geq 0)$$

where $\alpha(A) = \max_{k=1,\dots,n} \operatorname{Re} \lambda_k(A)$. Moreover, by (3.1),

$$\|\exp(At)\| \leq e^{\alpha(A)t} \sum_{k=0}^{n-1} \frac{g^k(A) t^k}{(k!)^{3/2}} \quad (t \geq 0). \tag{8.1}$$

Taking into account that the operator $\exp(-At)$ is the inverse one to $\exp(At)$ it is not hard to show that

$$\|\exp(At)h\| \geq \frac{e^{\beta(A)t}\|h\|}{\sum_{k=0}^{n-1} g^k(A)t^k(k!)^{-1}\gamma_{n,k}} \quad (t \geq 0, h \in \mathbb{C}^n),$$

where $\beta(A) = \min_{k=1,\dots,n} \operatorname{Re} \lambda_k(A)$. Therefore by (3.1),

$$\|\exp(At)h\| \geq \frac{e^{\beta(A)t}\|h\|}{\sum_{k=0}^{n-1} g^k(A)(k!)^{-3/2}\, t^k} \quad (t \geq 0). \tag{8.2}$$

Moreover, if A is a diagonalizable $n \times n$-matrix, then due to Theorem 2.7.1 we conclude that

$$\|e^{-At}\| \leq \gamma(A)e^{-\beta(A)t} \ (t \geq 0).$$

Hence,

$$\|e^{At}h\| \geq \frac{\|h\|e^{\beta(A)t}}{\gamma(A)} \quad (t \geq 0, h \in \mathbb{C}^n).$$

2.8.2 Perturbations of matrix exponentials

Let $A, \tilde{A} \in \mathbb{C}^{n\times n}$ and $E = \tilde{A} - A$. To investigate perturbations of matrix exponentials one can use the equation

$$e^{\tilde{A}t} - e^{At} = \int_0^t e^{\tilde{A}(t-s)}Ee^{As}ds \tag{8.3}$$

and estimate (8.1). Here we investigate perturbations in the case when $\|\tilde{A}E - EA\|$ is small enough. We will say that A is stable (Hurwitzian), if $\alpha(A) < 0$. Assume that A is stable and put

$$u(A) = \int_0^\infty \|e^{At}\|dt \quad \text{and} \quad v_A = \int_0^\infty t\|e^{At}\|dt.$$

Theorem 2.8.1. *Let A be stable, and*

$$\|\tilde{A}E - EA\|v_A < 1. \tag{8.4}$$

Then $\tilde{A}$ is also stable. Moreover,

$$u(\tilde{A}) \leq \frac{u(A) + v_A\|E\|}{1 - v_A\|\tilde{A}E - EA\|} \tag{8.5}$$

and

$$\int_0^\infty \|e^{\tilde{A}t} - e^{At}\|dt \leq \|E\|v_A + \frac{\|\tilde{A}E - EA\|v_A(u(A) + v_A\|E\|)}{1 - v_A\|\tilde{A}E - EA\|}. \tag{8.6}$$

This theorem is proved in the next subsection.

Furthermore, by (8.1) we obtain $u(A) \le u_0(A)$ and $v_A \le \hat{v}_A$, where

$$u_0(A) := \sum_{k=0}^{n-1} \frac{g^k(A)}{|\alpha(A)|^{k+1}(k!)^{1/2}} \quad \text{and} \quad \hat{v}_A := \sum_{k=0}^{n-1} \frac{(k+1)g^k(A)}{|\alpha(A)|^{k+2}(k!)^{1/2}}.$$

Thus, Theorem 2.8.1 implies

Corollary 2.8.2. *Let A be stable and $\|\tilde{A}E - EA\|\hat{v}_A < 1$. Then $\tilde{A}$ is also stable. Moreover,*

$$u(\tilde{A}) \le \frac{u_0(A) + \hat{v}_A\|E\|}{1 - \hat{v}_A\|\tilde{A}E - EA\|}$$

and

$$\int_0^\infty \|e^{\tilde{A}t} - e^{At}\|dt \le \|E\|\hat{v}_A + \frac{\|\tilde{A}E - EA\|\hat{v}_A(u_0(A) + \hat{v}_A\|E\|)}{1 - \hat{v}_A\|\tilde{A}E - EA\|}.$$

2.8.3 Proof of Theorem 2.8.1

We use the following result, let $f(t), c(t)$ and $h(t)$ be matrix functions defined on $[0, b]$ $(0 < b < \infty)$. Besides, f and h are differentiable and c integrable. Then

$$\int_0^b f(t)c(t)h(t)dt = f(b)j(b)h(b) - \int_0^b (f'(t)j(t)h(t) + f(t)j(t)h'(t))dt$$

with

$$j(t) = \int_0^t c(s)ds.$$

For the proof see Lemma 8.2.1 below. By that result

$$e^{\tilde{A}t} - e^{At} = \int_0^t e^{A(t-s)}Ee^{As}ds = Ete^{At} + \int_0^t e^{\tilde{A}(t-s)}[\tilde{A}E - EA]se^{As}ds.$$

Hence,

$$\int_0^\infty \|e^{\tilde{A}t} - e^{At}\|dt \le \int_0^\infty \|Ete^{At}\|dt + \int_0^\infty \int_0^t \|e^{\tilde{A}(t-s)}\|\|\tilde{A}E - EA\|\|se^{As}\|ds\, dt.$$

But

$$\int_0^\infty \int_0^t \|e^{\tilde{A}(t-s)}\|s\|e^{As}\|ds\, dt = \int_0^\infty \int_s^\infty \|e^{\tilde{A}(t-s)}\|s\|e^{As}\|dt\, ds$$
$$= \int_0^\infty s\|e^{As}\|ds \int_0^\infty \|e^{\tilde{A}t}\|dt = v_A u(\tilde{A}).$$

Thus

$$\int_0^\infty \|e^{\tilde{A}t} - e^{At}\| dt \le \|E\| v_A + \|\tilde{A}E - EA\| v_A u(\tilde{A}). \tag{8.7}$$

Hence,

$$u(\tilde{A}) \le u(A) + \|E\| v_A + \|\tilde{A}E - EA\| v_A u(\tilde{A}).$$

So according to (8.4), we get (8.5). Furthermore, due to (8.7) and (8.5) we get (8.6) as claimed. □

2.9 Matrices with non-negative off-diagonals

In this section it is assumed that $A = (a_{ij})_{j,k=1}^n$ is a real matrix with

$$a_{ij} \ge 0 \text{ for } i \ne j. \tag{9.1}$$

Put

$$a = \min_{j=1,\dots,n} a_{jj} \quad \text{and} \quad b = \max_{j=1,\dots,n} a_{jj}.$$

For a scalar function $f(\lambda)$ write

$$\alpha_k(f, A) := \inf_{a \le x \le b} \frac{f^{(k)}(x)}{k!} \quad \text{and} \quad \beta_k(f, A) := \sup_{a \le x \le b} \frac{f^{(k)}(x)}{k!} \quad (k = 0, 1, 2, \dots),$$

assuming that the derivatives exist.

Let $W = A - \operatorname{diag}(a_{jj})$ be the off diagonal part of A.

Theorem 2.9.1. *Let condition (9.1) hold and $f(\lambda)$ be holomorphic on a neighborhood of a Jordan set, whose boundary C has the property*

$$|z - a_{jj}| > \sum_{k=1, k \ne j}^n a_{jk}$$

for all $z \in C$ and $j = 1, \dots, n$. In addition, let f be real on $[a, b]$. Then the following inequalities are valid:

$$f(A) \ge \sum_{k=0}^\infty \alpha_k(f, A) W^k, \tag{9.2}$$

provided

$$r_s(W) \lim_{k \to \infty} \sqrt[k]{|\alpha_k(f, A)|} < 1,$$

and

$$f(A) \le \sum_{k=0}^\infty \beta_k(f, A) W^k, \tag{9.3}$$

provided,

$$r_s(W) \lim_{k \to \infty} \sqrt[k]{|\beta_k(f, A)|} < 1.$$

In particular, if $\alpha_k(f, A) \ge 0$ $(k = 0, 1, \dots)$, then matrix $f(A)$ has non-negative entries.

Proof. By (6.2) and (6.3),

$$f(A) = f(S) + \sum_{k=1}^{\infty} M_k,$$

where

$$M_k = \sum_{j_1=1}^{n} \hat{Q}_{j_1} W \sum_{j_2=1}^{n} \hat{Q}_{j_2} W \dots W \sum_{j_{k+1}=1}^{n} \hat{Q}_{j_{k+1}} J_{j_1 j_2 \dots j_{k+1}}.$$

Here

$$J_{j_1 \dots j_{k+1}} = \frac{(-1)^{k+1}}{2\pi i} \int_C \frac{f(\lambda) d\lambda}{(b_{j_1} - \lambda) \dots (b_{j_{k+1}} - \lambda)} \quad (b_j = a_{jj}).$$

Since S is real, Lemma 1.5.2 from [31] gives us the inequalities

$$\alpha_k(f, A) \le J_{j_1 \dots j_{k+1}} \le \beta_k(f, A).$$

Hence,

$$M_k \ge \alpha_k(f, A) \sum_{j_1=1}^{n} \hat{Q}_{j_1} W \sum_{j_2=1}^{n} \hat{Q}_{j_2} W \dots W \sum_{j_{k+1}=1}^{n} \hat{Q}_{j_{k+1}} = \alpha_k(f, A) W^k.$$

Similarly, $M_k \le \beta_k(f, A) W^k$. This implies the required result. □

2.10 Comments

One of the first estimates for the norm of a regular matrix-valued function was established by I.M. Gel'fand and G.E. Shilov [19] in connection with their investigations of partial differential equations, but that estimate is not sharp; it is not attained for any matrix. The problem of obtaining a sharp estimate for the norm of a matrix-valued function has been repeatedly discussed in the literature, cf. [14]. In the late 1970s, the author obtained a sharp estimate for a matrix-valued function regular on the convex hull of the spectrum, cf. [21] and references therein. It is attained in the case of normal matrices. Later, that estimate was extended to various classes of non-selfadjoint operators, such as Hilbert–Schmidt operators, quasi-Hermitian operators (i.e., linear operators with completely continuous imaginary components), quasiunitary operators (i.e., operators represented as a sum of a unitary operator and a compact one), etc. For more details see [31, 56] and references given therein.

The material of this chapter is taken from the papers [51, 53, 56] and the monograph [31].

About the relevant results on matrix-valued functions and perturbations of matrices see the well-known books [9, 74] and [90].

Chapter 3

General Linear Systems

This chapter is devoted to general linear systems including the Bohl–Perron principle.

3.1 Description of the problem

Let $\eta < \infty$ be a positive constant, and $R(t, \tau)$ be a real $n \times n$-matrix-valued function defined on $[0, \infty) \times [0, \eta\,]$, which is piece-wise continuous in t for each τ, whose entries are left-continuous and have bounded variations in τ. From the theory of functions of a bounded variation (see for instance [73]), it is well known that for a fixed t, $R(t, \tau)$ can be represented as $R(t, \tau) = R_1(t, \tau) + R_2(t, \tau) + R_3(t, \tau)$, where $R_1(t, \tau)$ is a saltus function of bounded variation with at most countably many jumps on $[0, \eta\,]$, $R_2(t, \tau)$ is an absolutely continuous function and $R_3(t, \tau)$ is either zero or a singular function of bounded variation, i.e., $R_3(t, \tau)$ is non-constant, continuous and has the derivative in τ which is equal to zero almost everywhere on $[0, \eta\,]$. If $R_3(t, .)$ is not zero identically, then the expression

$$\int_0^\eta d_\tau R_3(t, \tau) f(t - \tau)$$

for a continuous function f, cannot be transformed to a Lebesgue integral or to a series. For a concrete example see [73, p. 457]. So *in the sequel it is assumed that* $R_3(t, \tau)$ *is zero identically.*

Consider $R_1(t, \tau)$ and let for a given $t > 0$, $h_1(t) < h_2(t) < \cdots$ be those points in $[0, \eta\,)$, where at least one entry of $R_1(t, .)$ has a jump. Define

$$A_k(t) = R_1(t, h_k(t) + 0) - R_1(t, h_k(t)).$$

Then

$$\int_0^\eta dR_1(t, \tau) f(t - \tau) = \sum_{k=1}^\infty A_k(t) f(t - h_k(t))$$

for any continuous function f.

Define $A(t,s)$ by

$$R_2(t,\tau) = \int_0^\tau A(t,s)ds, \quad 0 \le \tau \le \eta\,.$$

Then

$$\int_0^\eta dR_2(t,\tau)f(t-\tau) = \int_0^\eta A(t,s)f(t-s)ds.$$

Our main object in this chapter is the following problem in $\mathbb{C}^n$:

$$\dot{y}(t) = \int_0^\eta d_\tau R(t,\tau)y(t-\tau) \;\; (t \ge 0), \tag{1.1}$$

$$y(t) = \phi(t) \;\; (-\eta \le t \le 0) \tag{1.2}$$

for a given vector function $\phi(t) \in C(-\eta\,,0)$. Here and below $\dot{y}(t)$ is the right derivative of $y(t)$. The integral in (1.1) is understood as the Lebesgue–Stieltjes integral.

We need the corresponding non-homogeneous equation

$$\dot{x}(t) = \int_0^\eta d_\tau R(t,\tau)x(t-\tau) + f(t) \;\; (t > 0) \tag{1.3}$$

with a given locally integrable vector function $f(t)$ and the zero initial condition

$$x(t) = 0 \; (-\eta \le t \le 0). \tag{1.4}$$

A solution of problem (1.1), (1.2) is a continuous vector function $y(t)$ defined on $[-\eta\,,\infty)$ and satisfying

$$\begin{aligned} y(t) &= \phi(0) + \int_0^t \int_0^\eta d_\tau R(s,\tau)y(s-\tau)ds \;\; (t \ge 0), \\ y(t) &= \phi(t) \;\; (-\eta \le t \le 0). \end{aligned} \tag{1.5}$$

A solution of problem (1.3), (1.4) is a continuous vector function $x(t)$ defined on $[0,\infty)$ and satisfying

$$x(t) = \int_0^t [f(s) + \int_0^\eta d_\tau R(s,\tau)x(s-\tau)]ds \;\; (t \ge 0) \tag{1.6}$$

with condition (1.4). Below we prove that problems (1.1), (1.2) and (1.3), (1.4) have unique solutions. See also the well-known Theorem 6.1.1 from [71].

It is assumed that the variation of $R(t,\tau) = (r_{ij}(t,\tau))_{i,j=1}^n$ in τ is bounded on $[0,\infty)$:

$$v_{jk} := \sup_{t \ge 0} \operatorname{var}(r_{jk}(t,.)) < \infty. \tag{1.7}$$

For instance, consider the equation

$$\dot{y}(t) = \int_0^{\eta} A(t,s)y(t-s)ds + \sum_{k=0}^{m} A_k(t)y(t-\tau_k) \quad (t \geq 0;\ m < \infty), \tag{1.8}$$

where

$$0 = \tau_0 < \tau_1 < \cdots < \tau_m \leq \eta$$

are constants, $A_k(t)$ are piece-wise continuous matrices and $A(t,\tau)$ is integrable in τ on $[0,\eta]$. Then (1.8) can be written as (1.1). Besides, (1.7) holds, provided

$$\sup_{t\geq 0} \left(\int_0^{\eta} \|A(t,s)\|_n ds + \sum_{k=0}^{m} \|A_k(t)\|_n \right) < \infty. \tag{1.9}$$

In this chapter $\|z\|_n$ is the Euclidean norm of $z \in \mathbb{C}^n$ and $\|A\|_n$ is the spectral norm of a matrix A. In addition, $C(\chi) = C(\chi, \mathbb{C}^n)$ is the space of continuous functions defined on a set $\chi \in \mathbb{R}$ with values in $\mathbb{C}^n$ and the norm $\|w\|_{C(\chi)} = \sup_{t\in\chi} \|w(t)\|_n$.

Recall that in a finite-dimensional space all the norms are equivalent.

Let $y(t)$ be a solution of problem (1.1), (1.2). Then (1.1) is said to be *stable*, if there is a constant $c_0 \geq 1$, independent of ϕ, such that

$$\|y(t)\|_n \leq c_0 \|\phi\|_{C(-\eta,0)} \ (t \geq 0).$$

Equation (1.1) *is said to be asymptotically stable* if it is stable and $y(t) \to 0$ as $t \to \infty$. Equation (1.1) is *exponentially stable* if there are constants $c_0 \geq 1$ and $\nu > 0$, independent of ϕ, such that

$$\|y(t)\|_n \leq c_0 e^{-\nu t} \|\phi\|_{C(-\eta,0)} \ (t \geq 0).$$

3.2 Existence of solutions

Introduce the operator $E : C(-\eta,\infty) \to C(0,\infty)$ by

$$Eu(t) = \int_0^{\eta} d_\tau R(t,\tau)u(t-\tau) \ \ (t \geq 0;\ u \in C(-\eta,\infty)).$$

Lemma 3.2.1. *Let condition* (1.7) *hold. Then for any $T > 0$, there is a constant $V(R)$ independent of T, such that*

$$\|Eu\|_{C(0,T)} \leq V(R) \|u\|_{C(-\eta,T)}. \tag{2.1}$$

Proof. This result is due to Lemma 1.12.3. □

Lemma 3.2.2. *If condition* (1.7) *holds and a vector-valued function f is integrable on each finite segment, then problem* (1.3), (1.4) *has on* $[0,\infty)$ *a unique solution.*

Proof. By (1.6)

$$x = f_1 + Wx,$$

where

$$f_1(t) = \int_0^t f(s)ds + \int_0^t Ex(s)ds$$

and

$$Wx(t) = \int_0^t Ex(s)ds. \tag{2.2}$$

For a fixed $T < \infty$, by the previous lemma

$$\|Wx\|_{C(0,T)} \leq V(R) \int_0^T \|x\|_{C(0,s_1)} ds_1.$$

Hence,

$$\|W^2 x\|_{C(0,T)} \leq V^2(R) \int_0^T \int_0^{s_1} \|x\|_{C(0,s_2)} ds_2 ds_1.$$

Similarly,

$$\|W^k x\|_{C(0,T)} \leq V^k(R) \int_0^T \int_0^{s_1} \dots \int_0^{s_k} \|x\|_{C(0,s_k)} ds_k \dots ds_2 ds_1$$
$$\leq V^k(R) \|x\|_{C(0,T)} \frac{T^k}{k!}.$$

Thus the spectral radius of W equals zero and, consequently,

$$(I - W)^{-1} = \sum_{k=0}^{\infty} W^k.$$

Therefore,

$$x = (I - W)^{-1} f_1 = \sum_{k=0}^{\infty} W^k f_1. \tag{2.3}$$

This proves the lemma. □

Lemma 3.2.3. *If condition* (1.7) *holds, then problem* (1.1), (1.2) *has on* $[0, \infty)$ *a unique solution for an arbitrary* $\phi \in C(-\eta, 0)$.

Proof. Put $\hat{\phi}(t) = \phi(t)$ $(t \leq 0)$, $\hat{\phi}(t) = \phi(0)$ $(t \geq 0)$, and $x(t) = y(t) - \hat{\phi}(t)$ $(t \geq 0)$. Then problem (1.1), (1.2) takes the form (1.3), (1.4) with $f = E\hat{\phi}$. Now the previous lemma proves the result. □

3.3 Fundamental solutions

Let $G(t,\tau)$ $(\tau \geq 0; t \geq \tau - \eta)$ be a matrix-valued function satisfying the equation

$$G(t,\tau) = I + \int_\tau^t \int_0^\eta d_{s_1} R(s,s_1) G(s-s_1,\tau) ds \ \ (t \geq \tau), \tag{3.1}$$

with the condition

$$G(t,\tau) = 0 \ (\tau - \eta \leq t < \tau). \tag{3.2}$$

Then $G(t,\tau)$ is called the *fundamental solution of equation* (1.1).

Repeating the arguments of Lemma 3.2.2, we prove the existence and uniqueness of the fundamental solution.

Clearly, $G(\tau,\tau)$ is an absolutely continuous in t matrix-valued function, satisfying (1.1) and the conditions

$$G(\tau,\tau) = I, \ G(t,\tau) = 0 \ \ (t < \tau). \tag{3.3}$$

Any solution $x(t)$ of problem (1.3), (1.4) can be represented as

$$x(t) = \int_0^t G(t,\tau) f(\tau) d\tau. \tag{3.4}$$

This formula can be obtained by direct differentiation. About other proofs see [71, Section 6.2], [78]. In these books the solution representations for the homogeneous problem also can be found.

Formula (3.4) is called *the Variation of Constants formula.*

Now we are going to derive a representation for solutions of the homogeneous equation (1.1).

To this end put

$$z(t) = \begin{cases} y(t) - G(t,0)\phi(0) & \text{if } t \geq 0, \\ 0 & \text{if } -\eta \leq t < 0, \end{cases}$$

where $y(t)$ is a solution of problem (1.1), (1.2). Then

$$y(t) = \begin{cases} z(t) + G(t,0)\phi(0) & \text{if } t \geq 0, \\ \phi(t) & \text{if } -\eta \leq t < 0. \end{cases}$$

If we choose

$$\tilde{\phi}(t) = \begin{cases} 0 & \text{if } t \geq 0, \\ \phi(t) & \text{if } -\eta \leq t < 0, \end{cases} \tag{3.5}$$

then we can write

$$y(t) = z(t) + G(t,0)\phi(0) + \tilde{\phi}(t) \ \ (t \geq -\eta). \tag{3.6}$$

Substituting this equality into (1.1), we have

$$\dot{z}(t)+\frac{\partial}{\partial t}G(t,0)\phi(0)=\int_0^\eta d_\tau R(t,\tau)(z(t-\tau)+\tilde{\phi}(t-\tau)+G(t-\tau,0)\phi(0))\ \ (t\geq 0).$$

But

$$\frac{\partial}{\partial t}G(t,0)\phi(0)=\int_0^\eta d_\tau R(t,\tau)G(t-\tau,0)\phi(0)$$

so

$$\dot{z}(t)=\int_0^\eta d_\tau R(t,\tau)(z(t-\tau)+\tilde{\phi}(t-\tau))\ \ (t\geq 0).$$

By the Variation of Constants formula

$$z(t)=\int_0^t G(t,s)\int_0^\eta d_\tau R(s,\tau)\tilde{\phi}(s-\tau)ds.$$

Taking into account (3.6), we get the following result.

Corollary 3.3.1. *For a solution $y(t)$ of the homogeneous problem* (1.1), (1.2) *the representation*

$$y(t)=G(t,0)\phi(0)+\int_0^\eta\int_0^\tau G(t,s)d_\tau R(s,\tau)\phi(s-\tau)ds\ \ (t\geq 0) \tag{3.7}$$

is valid.

3.4 The generalized Bohl–Perron principle

Theorem 3.4.1. *If for any $f\in C(0,\infty)$, problem* (1.3), (1.4) *has a bounded on* $[0,\infty)$ *solution, and condition* (1.7) *holds, then equation (1.1) is exponentially stable.*

To prove this theorem we need the following lemma.

Lemma 3.4.2. *If for any $f\in C(0,\infty)$ a solution of problem* (1.3), (1.4) *is bounded, then any solution of problem* (1.1), (1.2) *is also bounded.*

Proof. Let $y(t)$ be a solution of problem (1.1), (1.2). Put

$$\hat{\phi}(t)=\begin{cases}\phi(0) & \text{if } t\geq 0,\\ \phi(t) & \text{if } -\eta\leq t<0,\end{cases}$$

and $x_0(t)=y(t)-\hat{\phi}(t)$. We can write $d\hat{\phi}(t)/dt=0\ \ (t\geq 0)$ and

$$\dot{x}_0(t)=\int_0^\eta d_\tau R(t,\tau)x_0(t-\tau)+\ \psi(t)\ \ (t>0),$$

where

$$\psi(t) = \int_0^\eta d_\tau R(t,\tau)\hat{\phi}(t-\tau).$$

Besides, (1.4) holds with $x(t) = x_0(t)$. Since $V(R) < \infty$, we have $\psi \in C(-\eta, \infty)$. Due to the hypothesis of this lemma, $x_0 \in C(0,\infty)$. Thus $y \in C(-\eta, \infty)$, as claimed. □

Lemma 3.4.3. *Let $a(s)$ be a continuous function defined on $[0,\eta]$ and condition (1.7) hold. Then for any $T > 0$, one has*

$$\left\| \int_0^\eta a(s) d_s R(t,s) f(t-s) \right\|_{C(0,T)} \leq V(R) \max_{0\leq s\leq \eta} |a(s)| \, \|f\|_{C(-\eta,T)},$$

where $V(R)$ is defined by (2.1) and independent of T.

Proof. Let $f_k(t)$ be the coordinate of $f(t)$. Then

$$\begin{aligned}\left| \int_0^\eta a(s) f_k(t-s) d_s r_{jk}(t,s) \right| &\leq \max_{0\leq s\leq \eta} |a(s) f_k(t-s)| \int_0^\eta |d_s r_{jk}(t,s)| \\ &\leq v_{jk} \max_{0\leq s\leq \eta} |a(s)| \max_{-\eta\leq t\leq T} |f_k(t)|.\end{aligned}$$

Now repeating the arguments of the proof of Lemma 1.12.3, we obtain the required result. □

Proof of Theorem 3.4.1. Substituting

$$y(t) = y_\epsilon(t) e^{-\epsilon t} \quad (\epsilon > 0) \tag{4.1}$$

with an $\epsilon > 0$ into (1.1), we obtain the equation

$$\dot{y}_\epsilon(t) = \epsilon y_\epsilon(t) + \int_0^\eta e^{\epsilon\tau} d_\tau R(t,\tau) y_\epsilon(t-\tau) \quad (t > 0). \tag{4.2}$$

Introduce in $C(0,\infty)$ the operator

$$\hat{G}f(t) := \int_0^t G(t,s) f(s) ds \quad (t \geq 0) \tag{4.3}$$

$(f \in C(0,\infty))$. By the hypothesis of the theorem, we have

$$x = \hat{G}f \in C(0,\infty) \text{ for any } f \in C(0,\infty). \tag{4.4}$$

So $\hat{G}$ is defined on the whole space $C(0,\infty)$. It is closed, since problem (1.3), (1.4) has a unique solution. Therefore $\hat{G}$ is bounded according to the Closed Graph theorem (see Section 1.3).

Consider now the equation

$$\dot{x}_\epsilon(t) = \epsilon x_\epsilon(t) + \int_0^\eta e^{\epsilon\tau} d_\tau R(t,\tau) x_\epsilon(t-\tau) + f(t) \tag{4.5}$$

with the zero initial condition. For solutions x and x_ϵ of (1.3) and (4.5), respectively, we have

$$\begin{aligned}\frac{d}{dt}(x - x_\epsilon)(t) &= \int_0^\eta d_\tau R(t,\tau)x(t-\tau) - \epsilon x_\epsilon(t) - \int_0^\eta e^{\epsilon\tau} d_\tau R(t,\tau)x_\epsilon(t-\tau) \\ &= \int_0^\eta d_\tau R(t,\tau)(x(t-\tau) - x_\epsilon(t-\tau)) + f_\epsilon(t),\end{aligned}$$

where

$$f_\epsilon(t) = -\epsilon x_\epsilon(t) + \int_0^\eta (1 - e^{\epsilon\tau}) d_\tau R(t,\tau)x_\epsilon(t-\tau). \tag{4.6}$$

Consequently,

$$x - x_\epsilon = \hat{G} f_\epsilon. \tag{4.7}$$

For brevity in this proof we put $\|.\|_{C(0,T)} = |.|_T$ for a finite $T > 0$. Then

$$|\hat{G}|_T \le \|\hat{G}\|_{C(0,\infty)}$$

and due to Lemma 3.4.3,

$$|f_\epsilon|_T \le |x_\epsilon|_T(\epsilon + V(R)(e^{\epsilon\eta} - 1)).$$

So

$$|x_\epsilon|_T \le |x|_T + |x_\epsilon|_T \|\hat{G}\|_{C(0,\infty)}(\epsilon + V(R)(e^{\epsilon\eta} - 1)).$$

Thus for a sufficiently small ϵ,

$$\begin{aligned}|x_\epsilon|_T &\le \frac{|x|_T}{1 - \|\hat{G}\|_{C(0,\infty)}(\epsilon + V(R)(e^{\epsilon\eta} - 1))} \\ &\le \frac{\|x\|_{C(0,\infty)}}{1 - \|\hat{G}\|_{C(0,\infty)}(\epsilon + V(R)(e^{\epsilon\eta} - 1))} < \infty.\end{aligned}$$

Now letting $T \to \infty$, we get $x_\epsilon \in C(0,\infty)$. Hence, by the previous lemma, a solution y_ϵ of (4.2) is bounded. Now (4.1) proves the exponential stability, as claimed. □

3.5 L^p-version of the Bohl–Perron principle

In this chapter $L^p(\chi) = L^p(\chi, \mathbb{C}^n)$ $(p \ge 1)$ is the space of functions defined on a set $\chi \subset \mathbb{R}$ with values in $\mathbb{C}^n$ and the finite norm

$$\|w\|_{L^p(\chi)} = \left[\int_\chi \|w(t)\|_n^2 dt\right]^{1/2} \quad (w \in L^p(\chi);\ 1 \le p < \infty),$$

and $\|w\|_{L^\infty(\chi)} = \text{vrai}\sup_{t\in\chi} \|u(t)\|_n$. Besides, $R(t,\tau)$ is the same as in Section 3.1. In particular, condition (1.7) holds.

Theorem 3.5.1. *If for a $p \geq 1$ and any $f \in L^p(0,\infty)$, the non-homogeneous problem (1.3), (1.4) has a solution $x \in L^p(0,\infty)$, and condition (1.7) holds, then equation (1.1) is exponentially stable.*

The proof of this theorem is divided into a series of lemmas presented in this section.

Note that the existence and uniqueness of solutions of (1.3) in the considered case is due to the above proved Lemma 3.2.1, since f is locally integrable.

Again put

$$Eu(t) = \int_0^\eta d_\tau R(t,\tau)u(t-\tau) \quad (t \geq 0) \tag{5.1}$$

considering that operator as the one acting from $L^p(-\eta\,,T)$ into $L^p(0,T)$ for all $T > 0$.

Lemma 3.5.2. *For any $p \geq 1$ and all $T > 0$, there is a constant $V(R)$ independent of T, such that*

$$\|Eu\|_{L^p(0,T)} \leq V(R)\|u\|_{L^p(-\eta\,,T)}. \tag{5.2}$$

Proof. This result is due to Corollary 1.12.4. □

Lemma 3.5.3. *For any $f \in L^p(0,\infty)$ $(p \geq 1)$, let a solution of the nonhomogeneous problem (1.3), (1.4) be in $L^p(0,\infty)$ and (1.7) hold. Then any solution of the homogeneous problem (1.1), (1.2) is in $L^p(-\eta\,,\infty)$.*

Proof. With a $\mu > 0$, put

$$v(t) = \begin{cases} e^{-\mu t}\phi(0) & \text{if } t \geq 0, \\ \phi(t) & \text{if } -\eta\ \leq t < 0. \end{cases}$$

Then $v \in L^p(-\eta\,,\infty)$ and therefore $Ev \in L^p(0,\infty)$ for all $p \geq 1$. Furthermore, substitute $y(t) = x(t) + v(t)$ into (1.1). Then we have problem (1.3), (1.4) with

$$f(t) = \mu e^{-\mu t}\phi(0) + (Ev)(t).$$

Clearly, $f \in L^p(0,\infty)$. According to the assumption of this lemma, the solution $x(t)$ of problem (1.3), (1.4) is in $L^p(0,\infty)$. Thus, $y = x + v \in L^p(0,\infty)$, as claimed. □

Lemma 3.5.4. *If condition (1.7) holds and a solution $y(t)$ of problem (1.1), (1.2) is in $L^p(0,\infty)$ for a $p \geq 1$, then the solution is bounded on $[0,\infty)$. Moreover, if $p < \infty$, then*

$$\|y\|^p_{C(0,\infty)} \leq pV(R)\|y\|^p_{L^p(-\eta\,,\infty)}$$

where $V(R)$ is defined by (5.2).

Proof. By (1.1) and Lemma 3.5.2,

$$\|\dot{y}\|_{L^p(0,\infty)} \le V(R)\|y\|_{L^p(-\eta,\infty)}.$$

For simplicity, in this proof put $\|y(t)\|_n = |y(t)|$. The case $p = 1$ is obvious, since,

$$|y(t)| = -\int_t^\infty \frac{d|y(t_1)|}{dt_1}dt_1 \le \int_t^\infty |\dot{y}(t_1)|dt_1 \le \|\dot{y}\|_{L^1} \le V(R)\|y\|_{L^1} \ (t \ge 0).$$

Assume that $1 < p < \infty$. Then by the Gólder inequality

$$|y(t)|^p = -\int_t^\infty \frac{d|y(t_1)|^p}{dt_1}dt_1 = -p\int_t^\infty |y(t_1)|^{p-1}\frac{d|y(t_1)|}{dt_1}dt_1$$
$$\le p\int_t^\infty |y(t)|^{p-1}|\dot{y}(t)|dt \le p\left[\int_t^\infty |y(t)|^{q(p-1)}dt\right]^{1/q}\left[\int_t^\infty |\dot{y}(t)|dt\right]^{1/p}$$

where $q = p/(p-1)$. Since $q(p-1) = p$, we get the inequalities

$$|y(t)|^p \le p\|y\|^{p-1}_{L^p(0,\infty)}\|\dot{y}\|_{L^p(0,\infty)} \le p\|y\|^p_{L^p(-\eta,\infty)}V(R) \ \ (t \ge 0),$$

as claimed. □

Lemma 3.5.5. *Let $a(s)$ be a continuous function defined on $[0,\eta\,]$ and condition* (1.7) *hold. Then for any $T > 0$ and $p \ge 1$, one has*

$$\left\|\int_0^\eta a(s)d_sR(t,s)f(t-s)\right\|_{L^p(-\eta,T)} \le V(R)\max_{0\le s\le\eta}|a(s)|\ \|f\|_{L^p(-\eta,T)},$$

where $V(R)$ is defined by (5.2) *and independent of T.*

The proof of this lemma is similar to the proof of Lemma 3.4.3.

Proof of Theorem 3.5.1. Substituting (4.1) into (1.1), we obtain equation (4.2). Define the operator $\hat{G}$ on $L^p(0,\infty)$ by expression (4.3). By the hypothesis of the theorem, $x = \hat{G}f \in L^p(0,\infty)$ for all $f \in L^p(0,\infty)$. So $\hat{G}$ is defined on the whole space $L^p(0,\infty)$. It is closed, since problem (1.3), (1.4) has a unique solution. Therefore $\hat{G}$ is bounded according to the above-mentioned Closed Graph theorem.

Consider now equation (4.5) with the zero initial condition. According to (1.3), we have equation (4.7), with f_ϵ defined by (4.6), where x and x_ϵ are solutions of (1.3) and (4.5), respectively. For brevity, in this proof put $\|.\|_{L^p(0,T)} = |.|_{p,T}$ for a finite $T > 0$. Then $|\hat{G}|_{p,T} \le \|\hat{G}\|_{L^p(0,\infty)}$. In addition, by (4.6) and Lemma 3.5.5, we can write

$$|f_\epsilon|_{p,T} \le |x_\epsilon|_{p,T}(\epsilon + V(R)(e^{\epsilon\eta} - 1)).$$

Hence (4.7) implies the inequality

$$|x_\epsilon|_{p,T} \le |x|_{p,T} + \|\hat{G}\|_{L^p(0,\infty)}|x_\epsilon|_{p,T}(\epsilon + V(R)(e^{\epsilon\eta} - 1)).$$

Consequently, for all sufficiently small ϵ,

$$|x_\epsilon|_{p,T} \le \frac{|x|_{p,T}}{1 - \|\hat{G}\|_{L^p(0,\infty)}(\epsilon + V(R)(e^{\epsilon\eta} - 1))} \le |x_\epsilon|_{p,T}$$
$$\le \frac{\|x\|_{L^p(0,\infty)}}{1 - \|\hat{G}\|_{L^p(0,\infty)}(\epsilon + V(R)(e^{\epsilon\eta} - 1))} < \infty.$$

Letting $T \to \infty$, we get $x_\epsilon \in L^p(0,\infty)$. Consequently, by Lemmas 3.5.3 and 3.5.4, the solution y_ϵ of (4.2) is bounded. Now (4.1) proves the exponential stability, as claimed. □

3.6 Equations with infinite delays

Consider in $\mathbb{C}^n$ the equation

$$\dot{y}(t) = \int_0^\infty d_\tau R(t,\tau) y(t-\tau) \ \ (t \ge 0), \tag{6.1}$$

where $R(t,\tau)$ is an $n \times n$-matrix-valued function defined on $[0,\infty)^2$, which is continuous in t for each $\tau \ge 0$ and has bounded variation in τ for each $t \ge 0$. The integral in (6.1) is understood as the improper vector Lebesgue–Stieltjes integral. Take the initial condition

$$y(t) = \phi(t) \ \ (-\infty < t \le 0) \tag{6.2}$$

for a given $\phi \in C(-\infty,0) \cap L^1(-\infty,0)$. Consider also the nonhomogeneous equation

$$\dot{x}(t) = \int_0^\infty d_\tau R(t,\tau) x(t-\tau) + f(t) \ \ (t \ge 0) \tag{6.3}$$

with a given vector function $f(t)$ and the zero initial condition

$$x(t) = 0 \ (-\infty \le t \le 0). \tag{6.4}$$

In space $L^p(-\infty,\infty)$ with a $p \ge 1$ introduce the operator

$$E_\infty u(t) = \int_0^\infty d_\tau R(t,\tau) u(t-\tau) \ \ (t \ge 0;\ u \in L^p(-\infty,\infty)).$$

It is assumed that the inequality

$$\|E_\infty u\|_{L^p(0,\infty)} \le V(R) \|u\|_{L^p(-\infty,\infty)} \tag{6.5}$$

holds for a constant $V(R)$ and all $u \in L^p(-\infty,\infty)$.

A solution of problem (6.1), (6.2) is defined as the one of problem (1.1), (1.2) wit $\eta = \infty$; a solution of problem (6.3), (6.4) is defined as the one of problem

(1.3), (1.4). If $f \in L^p(0,\infty)$, $p \geq 1$, the existence and uniqueness of solutions can be proved as in Lemma 3.2.2, since f is locally integrable.

For instance, consider the equation

$$\dot{y}(t) = \int_0^\infty A(t,\tau)y(t-\tau)d\tau + \sum_{k=0}^\infty A_k(t)y(t-\tau_k) \quad (t \geq 0), \tag{6.6}$$

where $0 = \tau_0 < \tau_1, \dots$ are constants, $A_k(t)$ are piece-wise continuous matrices and $A(t,\tau)$ is integrable in τ on $[0,\infty)$. Then it is not hard to check that (6.5) holds, if

$$\sup_{t\geq 0} \left(\int_0^\infty \|A(t,s)\|_n ds + \sum_{k=0}^\infty \|A_k(t)\|_n \right) < \infty. \tag{6.7}$$

According to (6.4) equation (6.3) can be written as

$$\dot{x}(t) = \int_0^t d_\tau R(t,\tau)x(t-\tau) + f(t) \quad (t \geq 0). \tag{6.8}$$

We will say that *equation* (6.1) *has the ϵ-property* in L^p, if the relation

$$\left\| \int_0^t (e^{\epsilon\tau} - 1) d_\tau R(t,\tau) f(t-\tau) \right\|_{L^p(0,\infty)} \to 0 \text{ as } \epsilon \to 0 \ (\epsilon > 0) \tag{6.9}$$

holds for any $f \in L^p(-\infty,\infty)$.

The exponential stability of equation (6.1) is defined as in Section 3.1 with $\eta = \infty$.

Theorem 3.6.1. *For a $p \geq 1$ and any $f \in L^p(0,\infty)$, let the nonhomogeneous problem* (6.3), (6.4) *have a solution $x \in L^p(0,\infty)$. If, in addition, equation* (6.1) *has the ϵ-property and condition* (6.5) *holds, then that equation is exponentially stable.*

The proof of this theorem is divided into a series of lemmas presented in the next section.

3.7 Proof of Theorem 3.6.1

Lemma 3.7.1. *Under the hypothesis of Theorem* 3.6.1, *any solution of the homogeneous problem* (6.1), (6.2) *is in $L^p(0,\infty)$.*

Proof. With a $\mu > 0$, put

$$v(t) = \begin{cases} e^{-\mu t}\phi(0) & \text{if } t \geq 0, \\ \phi(t) & \text{if } -\infty \leq t < 0. \end{cases}$$

Clearly,

$$\|\phi\|^p_{L^p(-\infty,0)} \le \|\phi\|_{L^1(-\infty,0)}\|\phi\|^{p-1}_{C(-\infty,0)}.$$

So $v \in L^p(-\infty,\infty)$. By (6.5) we have $E_\infty v \in L^p(0,\infty)$. Substitute $y = v + x$ into (6.1). Then we have problem (6.3), (6.4) with

$$f(t) = \mu e^{-\mu t}\phi(0) + E_\infty v(t).$$

By the condition of the lemma $x \in L^p(0,\infty)$. We thus get the required result. □

Lemma 3.7.2. *If a solution $y(t)$ of problem* (6.1), (6.2) *is in $L^p(0,\infty)$ $(p \ge 1)$, and condition* (6.5) *holds, then it is bounded on $[0,\infty)$. Moreover, if $p < \infty$, then the inequalities*

$$\|y\|^p_{C(0,\infty)} \le pV(R)\|y\|^{p-1}_{L^p(0,\infty)}\|y\|_{L^p(-\infty,\infty)} \le pV(R)\|y\|^p_{L^p(-\infty,\infty)}$$

are valid.

The proof of this lemma is similar to the proof of Lemma 3.5.4.

Proof of Theorem 3.6.1. Substituting

$$y_\epsilon(t) = e^{t\epsilon}y(t)\ (t \ge 0) \quad \text{and} \quad y_\epsilon(t) = \phi(t)\ (t \le 0)$$

with an $\epsilon > 0$ into (6.1), we obtain the equation

$$\dot y_\epsilon(t) = \epsilon y_\epsilon(t) + \int_0^\infty e^{\epsilon\tau} d_\tau R(t,\tau) y_\epsilon(t-\tau)\ \ (t > 0). \tag{7.1}$$

Again use the operator $\hat G : f \to x$, where x is a solution of problem (6.3), (6.4). In other words,

$$\hat G f(t) = \int_0^t G(t,s)f(s)ds\ \ (t \ge 0;\ f \in L^p(0,\infty)),$$

where G is the fundamental solution to (6.1). By the hypothesis of the theorem, we have

$$x = \hat G f \in L^p(0,\infty) \text{ for any } f \in L^p(0,\infty).$$

So $\hat G$ is defined on the whole space $L^p(0,\infty)$. It is closed, since problem (6.3), (6.4) has a unique solution. Therefore $\hat G$ is bounded according to the Closed Graph theorem (see Section 1.3).

Consider now the equation

$$\dot x_\epsilon(t) = \epsilon x_\epsilon(t) + \int_0^\infty e^{\epsilon\tau} d_\tau R(t,\tau)x_\epsilon(t-\tau) + f(t) \tag{7.2}$$

with the zero initial condition. For solutions x and x_ϵ of (6.3) and (7.2), respectively, we have

$$\frac{d}{dt}(x - x_\epsilon)(t) = \int_0^\infty d_\tau R(t,\tau)x(t-\tau) - \epsilon x_\epsilon(t) - \int_0^\infty e^{\epsilon\tau} d_\tau R(t,\tau)x_\epsilon(t-\tau)$$
$$= \int_0^\infty d_\tau R(t,\tau)(x(t-\tau) - x_\epsilon(t-\tau)) + f_\epsilon(t),$$

where

$$f_\epsilon(t) = -\epsilon x_\epsilon(t) + \int_0^\infty (1 - e^{\epsilon\tau}) d_\tau R(t,\tau) x_\epsilon(t-\tau). \tag{7.3}$$

Consequently,

$$x - x_\epsilon = \hat{G} f_\epsilon. \tag{7.4}$$

For brevity in this proof, for a fixed p we put $\|.\|_{L^p(0,T)} = |.|_T$ for a finite $T > 0$. Then $|\hat{G}|_T \le \|\hat{G}\|_{L^p(0,\infty)}$. In addition, by the ϵ-property

$$|f_\epsilon(t)|_T \le v(\epsilon)|x_\epsilon|_T$$

where $v(\epsilon) \to 0$ as $\epsilon \to 0$. So

$$|x - x_\epsilon|_T \le v(\epsilon)\|\hat{G}\|_{L^p(0,\infty)}|x_\epsilon|_T.$$

Thus for a sufficiently small ϵ,

$$|x_\epsilon|_T \le \frac{|x|_T}{1 - v(\epsilon)\|\hat{G}\|_{L^p(0,\infty)}} \le \frac{\|x\|_{L^p(0,\infty)}}{1 - v(\epsilon)\|\hat{G}\|_{L^p(0,\infty)}} < \infty.$$

Now letting $T \to \infty$, we get $x_\epsilon \in L^p(0,\infty)$. Hence, by the previous lemma, a solution y_ϵ of (7.2) is bounded. Now (4.1) proves the exponential stability, as claimed. □

3.8 Equations with continuous infinite delay

In this section we illustrate Theorem 3.6.1 in the case $p = 2$. To this end consider in $\mathbb{C}^n$ the equation

$$\dot{y}(t) = \int_0^\infty A(\tau)y(t-\tau)d\tau + \int_0^\infty K(t,\tau)y(t-\tau)d\tau \ \ (t \ge 0), \tag{8.1}$$

where $A(\tau)$ is a piece-wise continuous matrix-valued function defined on $[0,\infty)$ and $K(t,\tau)$ is a piece-wise continuous matrix-valued function defined on $[0,\infty)^2$. Besides,

$$\|A(s)\|_n \le Ce^{-\mu s} \ \text{ and } \ \|K(t,s)\|_n \le Ce^{-\mu(t+s)} \ \ (C, \mu = \text{const} > 0;\ t, s \ge 0). \tag{8.2}$$

Then, it is not hard to check that (8.1) has in L^2 the ϵ-property, and the operator $\tilde{K} : L^2(-\infty, \infty) \to L^2(0, \infty)$, defined by

$$\tilde{K}w(t) = \int_0^\infty K(t, \tau)w(t - \tau)d\tau$$

is bounded. To apply Theorem 3.6.1, consider the equation

$$\dot{x}(t) = \int_0^t A(\tau)x(t - \tau)d\tau + \int_0^t K(t, \tau)x(t - \tau)d\tau + f(t) \ \ (t \geq 0) \tag{8.3}$$

with $f \in L^2(0, \infty)$. To estimate the solutions of the latter equation, we need the equation

$$\dot{u}(t) = \int_0^t A(\tau)u(t - \tau)d\tau + h(t) \ \ (t \geq 0) \tag{8.4}$$

with $h \in L^2(0, \infty)$. Applying to (8.4) the Laplace transform, we have

$$z\hat{u}(z) = \hat{A}(z)\hat{u}(z) + \hat{h}(z)$$

where $\hat{A}(z), \hat{u}(z)$ and $\hat{h}(z)$ are the Laplace transforms of $A(t), u(t)$ and $h(t)$, respectively, and z is the dual variable. Then

$$\hat{u}(z) = (zI - \hat{A}(z))^{-1}\hat{h}(z).$$

It is assumed that det $(zI - \hat{A}(z))$ is a stable function, that is all its zeros are in the open left half-plane, and

$$\begin{aligned} \theta_0 &:= \sup_{\omega \in \mathbb{R}} \|(i\omega I - \hat{A}(i\omega))^{-1}\|_n \\ &< \frac{1}{\|\tilde{K}\|_{L^2(0,\infty)}}. \end{aligned} \tag{8.5}$$

Note that various estimates for θ_0 can be found in Section 2.3. By the Parseval equality we have $\|u\|_{L^2(0,\infty)} \leq \theta_0 \|h\|_{L^2(0,\infty)}$. By this inequality, from (8.3) we get

$$\begin{aligned} \|x\|_{L^2(0,\infty)} &\leq \theta_0 \|f + \tilde{K}x\|_{L^2(0,\infty)} \\ &\leq \theta_0 \|f\|_{L^2(0,\infty)} + \theta_0 \|\tilde{K}\|_{L^2(0,\infty)} \|x\|_{L^2(0,\infty)}. \end{aligned}$$

Hence (8.5) implies that $x \in L^2(0, \infty)$. Now by Theorem 3.6.1 we get the following result.

Corollary 3.8.1. *Let conditions* (8.2) *and* (8.5) *hold. Then* (8.1) *is exponentially stable.*

3.9 Comments

The classical theory of differential delay equations is presented in many excellent books, for instance see [67, 72, 77, 96].

Recall that the Bohl–Perron principle means that the homogeneous ordinary differential equation (ODE) $\dot{y} = A(t)y$ $(t \geq 0)$ with a variable $n \times n$-matrix $A(t)$, bounded on $[0, \infty)$ is exponentially stable, provided the nonhomogeneous ODE $\dot{x} = A(t)x + f(t)$ with the zero initial condition has a bounded solution for any bounded vector-valued function f, cf. [14]. In [70, Theorem 4.15] the Bohl–Perron principle was generalized to a class of retarded systems with $R(t, \tau) = r(t, \tau)I$, where $r(t, \tau)$ is a scalar function; besides the asymptotic (not exponential) stability was proved (see also the book [4]).

Theorems 3.4.1 and 3.5.1 is proved in [58], Theorem 3.6.1 appears in the paper [59] (see also [61]).

Chapter 4

Time-invariant Linear Systems with Delay

This chapter is devoted to time-invariant (autonomous) linear systems with delay. In the terms of the characteristic matrix-valued function we derive estimates for the L^p- and C-norms of fundamental solutions. By these estimates below we obtain stability conditions for linear time-variant and nonlinear systems. Moreover, these estimates enable us to establish bounds for the region of attraction of the stationary solutions of nonlinear systems.

4.1 Statement of the problem

Everywhere in this chapter $R_0(\tau) = (r_{jk}(\tau))_{j,k=1}^n$ is a real left-continuous $n \times n$-matrix-valued function defined on a finite segment $[0, \eta\,]$, whose entries have bounded variations.

Recall that R_0 can be represented as $R_0 = R_1 + R_2 + R_3$, where R_1 is a saltus function of bounded variation with at most countably many jumps on $[0, \eta]$, R_2 is an absolutely continuous function and R_3 is either zero or a singular function of bounded variation, i.e., R_3 is non-constant, continuous and has a derivative which is equal to zero almost everywhere on $[0, \eta\,]$ [73]. As it was mentioned in Section 3.1, if R_3 is not zero identically, then the expression

$$\int_0^\eta dR_3(\tau) f(t - \tau)$$

for a continuous function f, cannot be transformed to a Lebesgue integral or to a series.

Consider $R_1(t)$ and let $h_1 < h_2, \ldots$ be those points in $[0, \eta\,)$, where at least one entry of R_1 has a jump. Then for any continuous function f,

$$\int_0^\eta dR_1(\tau) f(t - \tau) = \sum_{k=1}^\infty A_k f(t - h_k)$$

where A_k are constant matrices. R_1 being a function of bounded variation is equivalent to the condition

$$\sum_{k=1}^{\infty} \|A_k\|_n < \infty.$$

Here and below in this chapter, $\|A\|_n$ is the spectral norm of a matrix A.

For R_2 we define $A(s)$ by

$$R_2(\tau) = \int_0^{\tau} A(s)ds \ \ (0 \le \tau \le \eta\,).$$

Then

$$\int_0^{\eta} dR_2(\tau)f(t-\tau) = \int_0^{\eta} A(s)f(t-s)ds.$$

Since the function $R_2(\tau)$ is of bounded variation, we have

$$\int_0^{\eta} \|A(s)\|_n ds < \infty.$$

Our main object in this chapter is the problem

$$\dot{y}(t) = \int_0^{\eta} dR_0(\tau)y(t-\tau) \ (t \ge 0), \tag{1.1}$$

$$y(t) = \phi(t) \text{ for } -\eta \le t \le 0, \tag{1.2}$$

where $\phi(t)$ is a given continuous vector-valued function defined on $[-\eta\,, 0]$. Recall that $\dot{y}(t) = dy(t)/dt, t > 0$, and $\dot{y}(0)$ means the right derivative of y at zero.

As it was explained above in this section, equation (1.1) can be written as

$$\dot{y}(t) = \int_0^{\eta} A(s)y(t-s)ds + \sum_{k=1}^{m} A_k y(t-h_k) \ \ (t \ge 0;\ m \le \infty) \tag{1.3}$$

where $0 \le h_1 < h_2 < \cdots < h_m < \eta$ are constants, A_k are constant matrices and $A(s)$ is integrable on $[0, \eta\,]$. For most situations it is sufficient to consider the special case, where R_0 has only a finite number of jumps: $m < \infty$. So *in the sequel it is assumed that R_0 has a finite number of jumps.*

A solution $y(t)$ of problem (1.1), (1.2) is a continuous function

$$y := [-\eta\,, \infty) \to \mathbb{C}^n,$$

such that

$$y(t) = \phi(0) + \int_0^t \int_0^{\eta} dR_0(\tau)y(s-\tau)ds, t \ge 0, \tag{1.4a}$$

and

$$y(t) = \phi(t) \ \ (-\eta \le t \le 0). \tag{1.4b}$$

Recall that

$$\mathrm{Var}(R_0) = (\mathrm{var}(r_{ij}))_{i,j=1}^n$$

and

$$\mathrm{var}(R_0) = \| \mathrm{Var}(R_0)\|_n. \tag{1.5}$$

In particular, for equation (1.3), if $h_1 = 0$ we put $R_1(0) = 0$ and

$$R_1(\tau) = A_1 \ (0 < \tau \le h_2), R_1(\tau) = \sum_{k=1}^{j} A_k \text{ for } h_j < \tau \le h_{j+1}$$

$(j = 2, \dots, m;\ h_{m+1} = \eta\,)$. If $h_1 > 0$ we put $R_1(\tau) = 0\ (0 \le \tau \le h_1)$ and

$$R_1(\tau) = \sum_{k=1}^{j} A_k \text{ for } h_j < \tau \le h_{j+1} \ \ (j = 1, \dots, m).$$

In addition,

$$R_0(\tau) = \int_0^\tau A(s)ds + R_1(\tau).$$

Let $\tilde a_{ij}(s)$ and $a_{ij}^{(k)}$ be the entries of $A(s)$ and A_k, respectively. Then for equation (1.3), each entry of R_0, satisfies the inequality

$$\mathrm{var}(r_{ij}) \le \int_0^\eta |\tilde a_{ij}(s)|ds + \sum_{k=0}^{m} |a_{ij}^{(k)}|. \tag{1.6}$$

In this chapter again $C(a,b) = C([a,b];\mathbb{C}^n)$, $L^p(a,b) = L^p([a,b];\mathbb{C}^n)$.

Furthermore, Lemma 1.12.1 implies

$$\sup_{0\le s\le t} \left\| \int_0^\eta dR_0(\tau)y(s-\tau)\right\|_n \le \sqrt{n}\ \mathrm{var}(R_0) \sup_{-\eta\,\le s\le t} \|y(s)\|_n. \tag{1.7}$$

Put

$$\hat y(t) := \sup_{0\le s\le t} \|y(s)\|_n.$$

Then from (1.4) we deduce that

$$\hat y(t) \le \|\phi\|_{C(-\eta\,,0)} + \sqrt{n}\,\mathrm{var}(R_0) \int_0^t \hat y(s)ds.$$

Since $\|y(s)\|_n \le \hat y(s)$, by the Gronwall lemma we arrive at the inequality

$$\|y(t)\|_n \le \|\phi\|_{C(-\eta\,,0)} e^{t\sqrt{n}\,\mathrm{var}(R_0)} \ (t \ge 0). \tag{1.8}$$

4.2 Application of the Laplace transform

We need the non-homogeneous problem

$$\dot{x}(t) = \int_0^\eta dR_0(\tau)x(t-\tau) + f(t) \ (t \geq 0), \tag{2.1}$$

with a locally integrable f and the zero initial condition

$$x(t) = 0 \text{ for } -\eta \leq t \leq 0. \tag{2.2}$$

A solution of problem (2.1), (2.2) is a continuous vector function $x(t)$ defined on $[0, \infty)$ and satisfying the equation

$$x(t) = \int_0^t \int_0^\eta dR_0(\tau)x(s-\tau)ds + \int_0^t f(s)ds \ (t \geq 0) \tag{2.3}$$

and condition (2.2). Hence,

$$\|x\|_{C(0,t)} \leq \int_0^t \|f(s)\|_n ds + \sqrt{n} \ \mathrm{var}(R_0) \int_0^t \|x\|_{C(0,s)} ds.$$

Now the Gronwall lemma implies the inequality

$$\|x(t)\|_n \leq \|x\|_{C(0,t)} \leq \int_0^t \|f(s)\|_n ds \ \exp[(t\sqrt{n} \ \mathrm{var}(R_0)].$$

Assume also that f satisfies the inequality

$$\|f(t)\|_n \leq c_0 \ \ (c_0 = \text{const}) \tag{2.4}$$

almost everywhere on $[0, \infty)$.

Taking into account (1.8) we can assert that the Laplace transforms

$$\tilde{y}(z) = \int_0^\infty e^{-zt} y(t)dt \quad \text{and} \quad \tilde{x}(z) = \int_0^\infty e^{-zt} x(t)dt$$

of solutions of problems (1.1), (1.2) and (2.1), (2.2), respectively, exist at least for $\mathrm{Re}\, z > \sqrt{n} \ \mathrm{var}(R_0)$ and the integrals converge absolutely in this half-plane. In addition, inequality (1.7) together with equation (1.1) show that also $\dot{y}(t)$ has a Laplace transform at least in $\mathrm{Re}\, z > \sqrt{n} \ \mathrm{var}(R_0)$ given by $z\tilde{y}(z) - \phi(0)$. Taking the Laplace transforms of both sides of equation (1.1), we get

$$\begin{aligned} z\tilde{y}(z) - \phi(0) &= \int_0^\infty e^{-zt} \int_0^\eta dR_0(\tau)y(t-\tau)dt \\ &= \int_0^\eta e^{-\tau z} dR_0(\tau) \left[\int_{-\tau}^0 e^{-zt} y(t)dt + \tilde{y}(z) \right]. \end{aligned}$$

Interchanging the Stieltjes integration with the improper Riemann integration is justified by the Fubini theorem; we thus get

$$K(z)\tilde{y}(z) = \phi(0) + \int_0^\eta e^{-\tau z} dR_0(\tau) \int_{-\tau}^0 e^{-zt}\phi(t)dt,$$

where

$$K(z) = zI - \int_0^\eta e^{-\tau z} dR_0(\tau). \tag{2.5}$$

The matrix-valued function $K(z)$ is called *the characteristic matrix-valued function* to equation (1.1) and det $K(z)$ is called *the characteristic determinant* of equation (1.1). A zero of the characteristic determinant det $K(z)$ is called *a characteristic value of* $K(.)$ and $\lambda \in \mathbb{C}$ is *a regular value* of $K(.)$ if det $K(\lambda) \neq 0$.

In the sequel it is assumed that all the characteristic values of $K(.)$ *are in the open left half-plane* C_-.

Applying the inverse Laplace transform, we obtain

$$y(t) = \frac{1}{2\pi}\int_{-\infty}^{\infty} e^{i\omega t} K^{-1}(i\omega) \left[\phi(0) + \int_0^\eta dR_0(\tau) \int_{-\tau}^0 e^{-i\omega(s+\tau)}\phi(s)ds\right] d\omega \tag{2.6}$$

for $t \geq 0$. Furthermore, apply the Laplace transform to problem (2.1), (2.2). Then we easily obtain

$$\tilde{x}(z) = K^{-1}(z)\tilde{f}(z) \tag{2.7}$$

for all regular z. Here $\tilde{f}(z)$ is the Laplace transform to f. Applying the inverse Laplace transform, we get the following equality:

$$x(t) = \int_0^t G(t-s)f(s)ds \ \ (t \geq 0), \tag{2.8}$$

where

$$G(t) = \frac{1}{2\pi}\int_{-\infty}^{\infty} e^{i\omega t} K^{-1}(i\omega)d\omega. \tag{2.9}$$

Clearly, a matrix-valued function $G(t)$ satisfies (1.1). Moreover,

$$G(0+) = I, \ G(t) = 0 \ (t < 0). \tag{2.10}$$

So $G(t)$ is the fundamental solution of equation (1.1). Formula (2.8) is the Variation of Constants formula to problem (2.1), (2.2). Note that for equation (1.3) we have

$$K(z) = zI - \int_0^\eta e^{-sz} A(s)ds - \sum_{k=0}^m e^{-h_k z} A_k. \tag{2.11}$$

4.3 Norms of characteristic matrix functions

Let A be an $n \times n$-matrix. Recall that the quantity

$$g(A) = \left[N_2^2(A) - \sum_{k=1}^{n} |\lambda_k(A)|^2 \right]^{1/2},$$

is introduced in Section 2.3. Here $\lambda_k(A), k = 1, \dots, n$ are the eigenvalues of A, counted with their multiplicities; $N_2(A) = (\text{Trace}\, AA^*)^{1/2}$ is the Frobenius (Hilbert–Schmidt norm) of A, A^* is adjoint to A. As it is shown in Section 2.3, the following relations are valid:

$$g^2(A) \le N_2^2(A) - |\,\text{Trace}\, A^2| \quad \text{and} \quad g^2(A) \le \frac{N_2^2(A - A^*)}{2} = 2N_2^2(A_I), \tag{3.1}$$

where $A_I = (A - A^*)/2i$. Moreover,

$$g(e^{it}A + zI) = g(A) \ \ (t \in \mathbb{R}; z \in \mathbb{C}). \tag{3.2}$$

If A is a normal matrix: $AA^* = A^*A$, then $g(A) = 0$. Put

$$B(z) = \int_0^{\eta} e^{-z\tau} dR_0(\tau) \ \ (z \in \mathbb{C}).$$

In particular, for equation (1.3) we have

$$B(z) = \int_0^{\eta} e^{-sz} A(s) ds + \sum_{k=0}^{m} e^{-h_k z} A_k \tag{3.3}$$

and

$$g(B(i\omega)) \le N_2(B(i\omega)) \le \int_0^{\eta} N_2(A(s)) ds + \sum_{k=0}^{m} N_2(A_k) \ \ (\omega \in \mathbb{R}). \tag{3.4}$$

Below, under various assumptions, we suggest sharper estimates for $g(B(i\omega))$.

According to Theorem 2.3.1, the inequality

$$\|A^{-1}\|_n \le \sum_{k=0}^{n-1} \frac{g^k(A)}{\sqrt{k!}\, d^{k+1}(A)}$$

is valid for any invertible matrix A, where $d(A)$ is the smallest modulus of eigenvalues of A.

Hence we arrive at the inequality

$$\|[K(z)]^{-1}\|_n \le \Gamma(K(z)) \ (z \in \mathbb{C}), \tag{3.5}$$

where

$$\Gamma(K(z)) = \sum_{k=0}^{n-1} \frac{g^k(B(z))}{\sqrt{k!}d^{k+1}(K(z))}$$

and $d(K(z))$ is the smallest modulus of eigenvalues of matrix $K(z)$ for a fixed z:

$$d(K(z)) = \min_{k=1,\dots,n} |\lambda_k(K(z))|.$$

If $B(z)$ is a normal matrix, then $g(B(z)) = 0$, and

$$\|[K(z)]^{-1}\|_n \le \frac{1}{d(K(z))}.$$

For example that inequality holds, if $K(z) = zI - A_0 e^{-z\eta}$, where A_0 is a Hermitian matrix.

Let

$$\theta(K) := \sup_{-2\ \mathrm{var}(R_0) \le \omega \le 2\ \mathrm{var}(R_0)} \|K^{-1}(i\omega)\|_n.$$

Lemma 4.3.1. *The equality*

$$\sup_{-\infty \le \omega \le \infty} \|K^{-1}(i\omega)\|_n = \theta(K)$$

is valid.

Proof. We have

$$K(0) = \int_0^\eta dR_0(s) = \int_0^\eta dR_0(s) = R_0(\eta) - R_0(0).$$

So

$$\|K(0)\|_n = \|R_0(\eta) - R_0(0)\|_n \le \mathrm{var}(R_0),$$

and therefore,

$$\|K^{-1}(0)\|_n \ge \frac{1}{\mathrm{var}(R_0)}.$$

Here and below in this chapter, $\|v\|_n$ is the Euclidean norm of $v \in \mathbb{C}^n$. Simple calculations show that

$$\left\| \int_0^\eta e^{-i\omega} dR_0(\tau) \right\|_n \le \mathrm{var}(R_0) \quad (\omega \in \mathbb{R})$$

and

$$\|K(i\omega)v\|_n \ge (|\omega| - \mathrm{var}(R_0))\|v\|_n \ge \mathrm{var}(R_0)\|v\|_n$$
$$(\omega \in \mathbb{R}; |\omega| \ge 2\,\mathrm{var}(R_0);\ v \in \mathbb{C}^n).$$

So

$$\|K^{-1}(i\omega)\|_n \le \frac{1}{\mathrm{var}(R_0)} \le \|K^{-1}(0)\|_n \ (|\omega| \ge 2\,\mathrm{var}(R_0)).$$

Thus the maximum of $\|K^{-1}(i\omega)\|_n$ is attained inside the segment

$$[-2\ \mathrm{var}(R_0), 2\ \mathrm{var}(R_0)],$$

as claimed. □

By (3.4) and the previous lemma we have the inequality $\theta(K) \le \hat{\theta}(K)$, where

$$\hat{\theta}(K) := \sup_{|\omega|\le 2\,\mathrm{var}(R_0)} \Gamma(K(i\omega)).$$

Let

$$g_B := \sup_{\omega\in[-2\,\mathrm{var}(R_0),2\,\mathrm{var}(R_0)]} g(B(i\omega))$$

and

$$d_K := \inf_{\omega\in[-2\,\mathrm{var}(R_0),2\,\mathrm{var}(R_0)]} d(K(i\omega)).$$

Then we obtain the following result.

Corollary 4.3.2. *The inequalities*

$$\theta(K) \le \hat{\theta}(K) \le \Gamma_0(K), \tag{3.6}$$

are true, where

$$\Gamma_0(K) := \sum_{k=0}^{n-1} \frac{g_B^k}{\sqrt{k!}d_K^{k+1}}.$$

4.4 Norms of fundamental solutions of time-invariant systems

Recall that all the characteristic values of $K(.)$ are in the open left half-plane C_-. The following result directly follows from (2.6): let $y(t)$ be a solution of the time-invariant homogeneous problem (1.1), (1.2). Then due to (2.6)

$$y(t) = G(t)\phi(0) + \int_0^\eta \int_{-\tau}^0 G(t-\tau-s)dR_0(\tau)\phi(s)ds \ (t \ge 0). \tag{4.1}$$

Lemma 4.4.1. *Let $x(t)$ be a solution of problem (2.1), (2.2) with $f \in L^2(0,\infty)$. Then*

$$\|x\|_{L^2(0,\infty)} \le \theta(K)\|f\|_{L^2(0,\infty)}.$$

Proof. The result is due to (2.8), Lemma 4.3.1 and the Parseval equality. □

The previous lemma and Theorem 3.5.1 yield the following result.

Corollary 4.4.2. *Let all the characteristic values of $K(.)$ be in C_-. Then equation (1.1) is exponentially stable.*

Lemma 4.4.3. *The inequalities*

$$\left\| \int_0^\eta dR_0(s) f(t-s) \right\|_{L^2(0,\infty)} \le \mathrm{var}(R_0) \|f\|_{L^2(-\eta\,,\infty)} \ (f \in L^2(-\eta\,,\infty)) \tag{4.2}$$

and

$$\left\| \int_0^\eta dR_0(s) f(t-s) \right\|_{C(0,\infty)} \le \sqrt{n}\, \mathrm{var}(R_0) \|f\|_{C(-\eta\,,\infty)} \ (f \in C(-\eta\,,\infty)) \tag{4.3}$$

are true for all $t \ge 0$.

This result is due to Lemma 1.12.1.

Thanks to the Parseval equality,

$$\|G\|^2_{L^2(0,\infty)} = \frac{1}{2\pi} \|K^{-1}(i\omega)\|^2_{L^2(-\infty,\infty)} := \frac{1}{2\pi} \int_{-\infty}^{\infty} \|K^{-1}(i\omega)\|_n^2 d\omega.$$

Calculations of such integrals is often a difficult task. Because of this, in the next lemma we suggest an estimate for $\|G\|_{L^2(0,\infty)}$. Let

$$W(K) := \sqrt{2\theta(K)[\mathrm{var}(R_0)\theta(K) + 1]}. \tag{4.4}$$

Lemma 4.4.4. *The inequality* $\|G\|_{L^2(0,\infty)} \le W(K)$ *is valid.*

Proof. Set $w(t) = G(t) - \psi(t)$, where

$$\psi(t) = \begin{cases} Ie^{-bt} & \text{if } t \ge 0, \\ 0 & \text{if } -\eta \le t < 0, \end{cases}$$

with a positive constant b, which should be deduced. Then

$$\dot w(t) = \int_0^\eta dR_0(s) G(t-s) + be^{-bt} I = \int_0^\eta dR_0(s) w(t-s) + f(t) \ (t \ge 0),$$

where

$$f(t) = \int_0^\eta dR_0(s)\, \psi(t-s) + be^{-bt} I.$$

By Lemma 4.4.3, we have

$$\begin{aligned} \|f\|_{L^2(0,\infty)} &\le (\mathrm{var}(R_0) \|\,\psi\|_{L^2(0,\infty)} + b) \|e^{-bt}\|_{L^2(0,\infty)} \\ &\le (\mathrm{var}(R_0) + b) \|e^{-bt}\|_{L^2(0,\infty)}. \end{aligned}$$

Take into account that

$$\|e^{-tb}\|_{L^2(0,\infty)} = \frac{1}{\sqrt{2b}}.$$

Then due to Lemma 4.4.1, we get

$$\|w\|_{L^2(0,\infty)} \le \frac{\theta(K)(\mathrm{var}(R_0)+b)}{\sqrt{2b}}.$$

Hence,

$$\|G\|_{L^2(0,\infty)} \le \|w\|_{L^2(0,\infty)} + \frac{1}{\sqrt{2b}} = \frac{1}{\sqrt{2b}}(1+\theta(K)(\mathrm{var}(R_0)+b)).$$

The minimum of the right-hand part is attained for

$$b = \frac{1+\theta(K)\,\mathrm{var}(R_0)}{\theta(K)}.$$

Thus, we arrive at the inequality

$$\|G\|_{L^2(0,\infty)} \le \sqrt{2\theta(K)(\theta(K)\,\mathrm{var}(R_0)+1)},$$

as claimed. □

By Corollary 4.3.2 we have the inequality

$$W(K) \le \hat{W}(K), \quad \text{where} \quad \hat{W}(K) := \sqrt{2\hat{\theta}(K)[\mathrm{var}(R_0)\hat{\theta}(K)+1]}. \tag{4.5}$$

In addition, Lemma 4.4.3 and (1.1) imply the following lemma.

Lemma 4.4.5. *The inequality* $\|\dot{G}\|_{L^2(0,\infty)} \le \|G\|_{L^2(0,\infty)}\ \mathrm{var}(R_0)$ *is valid.*

Now (4.5) yields the inequalities

$$\|\dot{G}\|_{L^2(0,\infty)} \le W(K)\,\mathrm{var}(R_0) \le \hat{W}(K)\,\mathrm{var}(R_0). \tag{4.6}$$

We need also the following simple result.

Lemma 4.4.6. *Let* $f \in L^2(0,\infty)$ *and* $\dot{f} \in L^2(0,\infty)$. *Then*

$$\|f\|^2_{C(0,\infty)} \le 2\|f\|_{L^2(0,\infty)}\|\dot{f}\|_{L^2(0,\infty)}.$$

Proof. Obviously,

$$\|f(t)\|_n^2 = -\int_t^\infty \frac{d\|f(s)\|_n^2}{ds}ds = -2\int_t^\infty \|f(s)\|_n \frac{d\|f(s)\|_n}{ds}ds.$$

Taking into account that

$$\left|\frac{d}{ds}\|f(s)\|_n\right| \le \|\dot{f}(s)\|_n,$$

we get the required result due to Schwarz's inequality. □

By the previous lemma and (4.6) at once we obtain the following result.

Theorem 4.4.7. *The inequality*

$$\|G\|^2_{C(0,\infty)} \le 2\|G\|^2_{L^2(0,\infty)} \operatorname{var}(R_0)$$

is valid, and therefore,

$$\|G\|_{C(0,\infty)} \le a_0(K), \tag{4.7}$$

where

$$a_0(K) := \sqrt{2\operatorname{var}(R_0)W(K)} = 2\sqrt{\operatorname{var}(R_0)\theta(K)[\operatorname{var}(R_0)\theta(K)+1]}.$$

Clearly,

$$a_0(K) \le 2(1+\operatorname{var}(R_0)\theta(K)), \tag{4.8}$$

and by (4.5) and (3.6),

$$a_0(K) \le \sqrt{2\operatorname{var}(R_0)\hat{W}(K)} = 2\sqrt{\operatorname{var}(R_0)\hat{\theta}(K)[\operatorname{var}(R_0)\hat{\theta}(K)+1]}. \tag{4.9}$$

Now we are going to estimate the L^1-norm of the fundamental solution. To this end consider a function $r(s)$ of bounded variation. Then

$$r(s) = r_+(s) - r_-(s),$$

where $r_+(s), r_-(s)$ are nondecreasing functions. For a continuous function q defined on $[0,\eta]$, let

$$\int_0^\eta q(s)|dr(s)| := \int_0^\eta q(s)dr_+(s) + \int_0^\eta q(s)dr_-(s).$$

In particular, let

$$\operatorname{vd}(r) := \int_0^\eta s|dr(s)|.$$

Furthermore, put

$$\operatorname{vd}(R_0) = \|(\operatorname{vd}(r_{jk}))_{j,k=1}^n\|_n.$$

Recall that r_{jk} are the entries of R_0. That is, vd (R_0) is the spectral norm of the matrix whose entries are vd (r_{jk}).

Lemma 4.4.8. *The inequality*

$$\left\|\int_0^\eta \tau dR_0(\tau)f(t-\tau)\right\|_{L^2(0,T)} \le \operatorname{vd}(R_0)\|f\|_{L^2(-\eta,T)} \ (T>0;\ f\in L^2(-\eta,T))$$

is true.

Proof. Put

$$E_1 f(t) = \int_0^\eta \tau dR_0(\tau) f(t-\tau) \quad (f(t) = (f_k(t))).$$

Then we obtain

$$\|E_1 f(t)\|^2_{L^2(0,T)} = \sum_{j=1}^n \int_0^T |(E_1 f)_j(t)|^2 dt,$$

where $(E_1 f)_j(t)$ denotes the coordinate of $(E_1 f)(t)$. But

$$\begin{aligned} |(E_1 f)_j(t)|^2 &= \left| \sum_{k=1}^n \int_0^\eta s f_k(t-s) dr_{jk}(s) \right|^2 \le \sum_{k=1}^n \left(\int_0^\eta s|f_k(t-s)||dr_{jk}(s)| \right)^2 \\ &= \sum_{k=1}^n \int_0^\eta s|f_k(t-s)||dr_{jk}(s)| \sum_{i=1}^n \int_0^\eta s_1 |f_i(t-s_1)||dr_{ik}(s_1)|. \end{aligned}$$

Hence

$$\int_0^T |(E_1 f)_j(t)|^2 dt \le \int_0^\eta \int_0^\eta \sum_{i=1}^n \sum_{k=1}^n s|dr_{jk}(s)| s_1 |dr_{ji}(s_1)| \int_0^T |f_k(t-s) f_i(t-s_1)| dt.$$

By the Schwarz inequality

$$\begin{aligned} \left(\int_0^T |f_k(t-s) f_i(t-s_1)| dt \right)^2 &\le \int_0^T |f_k(t-s)|^2 dt \int_0^T |f_i(t-s_1)|^2 dt \\ &\le \int_{-\eta}^T |f_k(t)|^2 dt \int_{-\eta}^T |f_i(t)|^2 dt. \end{aligned}$$

Thus

$$\begin{aligned} \int_0^T |(E_1 f)_j(t)|^2 dt &\le \sum_{i=1}^n \sum_{k=1}^n \mathrm{vd}(r_{jk})\, \mathrm{vd}(r_{ik}) \|f_k\|_{L^2(-\eta,T)} \|f_i\|_{L^2(-\eta,T)} \\ &= \left(\sum_{k=1}^n \mathrm{vd}(r_{jk}) \|f_k\|_{L^2(-\eta,T)} \right)^2. \end{aligned}$$

Hence

$$\begin{aligned} \sum_{j=1}^n \int_0^T |(E_1 f)_j(t)|^2 dt &\le \sum_{j=1}^n \left(\sum_{k=1}^n \mathrm{vd}(r_{jk}) \|f_k\|_{L^2(-\eta,T)} \right)^2 \\ &= \left\| (\mathrm{vd}\,(r_{jk}))_{j,k=1}^n \nu_2 \right\|_n \le \mathrm{vd}\,(R_0) \|\nu_2\|_n, \end{aligned}$$

where

$$\nu_2 = \left(\|f_k\|_{L^2(-\eta,T)} \right)_{k=1}^n.$$

But $\|\nu_2\|_n = \|f\|_{L^2(-\eta,T)}$. This proves the lemma. □

Clearly

$$\mathrm{vd}(R_0) \le \eta \ \mathrm{var}(R_0).$$

For equation (1.3) we easily obtain

$$\mathrm{vd}(R_0) \le \int_0^{\eta} s\|A(s)\|_n ds + \sum_{k=0}^{m} h_k \|A_k\|_n. \tag{4.10}$$

We need the following technical lemma.

Lemma 4.4.9. *Let equation* (1.1) *be asymptotically stable. Then the function* $Y(t) := tG(t)$ *satisfies the inequality*

$$\|Y\|_{L^2(0,\infty)} \le \theta(K)(1 + \mathrm{vd}(R_0))\|G\|_{L^2(0,\infty)}.$$

Proof. By (1.1)

$$\begin{aligned}\dot{Y}(t) &= t\dot{G}(t) + G(t) = t\int_0^{\eta} dR_0(\tau)G(t-\tau) + G(t)\\ &= \int_0^{\eta} dR_0(\tau)(t-\tau)G(t-\tau) + \int_0^{\eta} \tau dR_0(\tau)G(t-\tau) + G(t).\end{aligned}$$

Thus,

$$\dot{Y}(t) = \int_0^{\eta} dR_0(\tau)Y(t-\tau) + F(t)$$

where

$$F(t) = \int_0^{\eta} \tau R(d\tau)G(t-\tau) + G(t).$$

Hence,

$$Y(t) = \int_0^{t} G(t-t_1)F(t_1)dt_1.$$

By Lemma 4.4.1, $\|Y\|_{L^2(0,\infty)} \le \theta(K)\|F\|_{L^2(0,\infty)}$. But due to the previous lemma

$$\|F\|_{L^2(0,\infty)} \le \|G\|_{L^2(0,\infty)}(1 + \mathrm{vd}(R_0)).$$

We thus have established the required result. □

Now we are in a position to formulate and prove the main result of the section.

Theorem 4.4.10. *The fundamental solution G to equation* (1.1) *satisfies the inequality*

$$\|G\|_{L^1(0,\infty)} \le \|G\|_{L^2(0\infty)}\sqrt{\pi\theta(K)(1 + \mathrm{vd}(R_0))} \tag{4.11}$$

and therefore,

$$\|G\|_{L^1(0,\infty)} \le W(K)\sqrt{\pi\theta(K)(1 + \mathrm{vd}(R_0))}. \tag{4.12}$$

Proof. Let us apply the Karlson inequality

$$\left(\int_0^\infty |f(t)|dt\right)^4 \leq \pi^2 \int_0^\infty f^2(t)dt \int_0^\infty t^2 f^2(t)dt$$

for a real scalar-valued $f \in L^2[0,\infty)$ with the property $tf(t) \in L^2[0,\infty)$, cf. [93, Chapter VIII]. By this inequality

$$\|G\|^4_{L^1(0,\infty)} \leq \pi^2 \|Y\|^2_{L^2(0,\infty)} \|G\|^2_{L^2(0,\infty)}.$$

Now the previous lemma yields the required result. □

Recall that $\theta(K)$ and $W(K)$ can be estimated by (3.6).

4.5 Systems with scalar delay-distributions

The aim of this section is to evaluate $g(B(z))$ for the system

$$\dot{y}(t) = \sum_{k=1}^m A_k \int_0^\eta y(t-s)d\mu_k(s), \tag{5.1}$$

where A_k are constant matrices and $\mu_k(s)$ are scalar nondecreasing functions defined on $[0,\eta]$.

In this case the characteristic matrix function is

$$K(z) = zI - \sum_{k=1}^m A_k \int_0^\eta e^{-zs} d\mu_k(s).$$

Simple calculations show that

$$\operatorname{var}(R_0) \leq \sum_{k=1}^m \|A_k\|_n \operatorname{var}(\mu_k) \quad \text{and} \quad \operatorname{vd}(R_0) \leq \sum_{k=1}^m \|A_k\|_n \int_0^\eta s\, d\mu_k(s).$$

In addition,

$$g(K(i\omega)) = g(B(i\omega)) \leq \sum_{k=1}^m N_2(A_k) \operatorname{var}(\mu_k) \ \ (\omega \in \mathbb{R}).$$

For instance, the equation

$$\dot{y}(t) = \sum_{k=1}^m A_k y(t-h_k) \tag{5.2}$$

can take the form (5.1). In the considered case

$$K(z) = zI - \sum_{k=1}^m A_k e^{-h_k z}, \quad \operatorname{var}(R_0) \leq \sum_{k=1}^m \|A_k\|_n$$

and

$$\mathrm{vd}(R_0) \leq \sum_{k=1}^{m} h_k \|A_k\|_n, \quad \text{and} \quad g(B(i\omega)) \leq \sum_{k=1}^{m} N_2(A_k) \ \ (\omega \in \mathbb{R}).$$

Under additional conditions, the latter estimate can be improved. For example, if

$$K(z) = zI - A_1 e^{-h_k z} - A_2 e^{-h_2 z}, \tag{5.3}$$

then due to (3.1) and (3.2), for all $\omega \in \mathbb{R}$, we obtain

$$\begin{aligned} g(B(i\omega)) &= g(e^{i\omega h_1} B(i\omega)) = g(A_1 + A_2 e^{(h_1-h_2)i\omega}) \\ &\leq \frac{1}{\sqrt{2}} N_2(A_1 - A_1^* + A_2 e^{(h_1-h_2)i\omega} - A_2^* e^{-(h_1-h_2)i\omega}) \end{aligned}$$

and, consequently,

$$g(B((i\omega)) \leq \frac{1}{\sqrt{2}} N_2(A_1 - A_1^*) + \sqrt{2} N_2(A_2). \tag{5.4}$$

Similarly, we get

$$g(B((i\omega)) \leq \sqrt{2} N_2(A_1) + \frac{1}{\sqrt{2}} N_2(A_2 - A_2^*) \ \ (\omega \in \mathbb{R}). \tag{5.5}$$

4.6 Scalar first-order autonomous equations

In this section we are going to estimate the fundamental solutions of some scalar equations. These estimates give us lower bounds for the quantity d_K. Recall that this quantity is introduced in Section 4.3. The obtained estimates will be applied in the next sections.

Consider a nondecreasing scalar function $\mu(s)$ ($s \in [0, \eta\,]$), and put

$$k(z) = z + \int_0^{\eta} \exp(-zs) d\mu(s) \ (z \in \mathbb{C}). \tag{6.1}$$

Obviously, $k(z)$ is the characteristic function of the scalar equation

$$\dot{y} + \int_0^{\eta} y(t-s) d\mu(s) = 0. \tag{6.2}$$

Lemma 4.6.1. *The equality*

$$\inf_{-2\,\mathrm{var}(\mu) \leq \omega \leq 2\,\mathrm{var}(\mu)} |k(i\omega)| = \inf_{\omega \in \mathbb{R}} |k(i\omega)|$$

is valid.

Proof. Clearly, $k(0) = \mathrm{var}(\mu)$ and

$$|k(i\omega)| \geq |\omega| - \mathrm{var}(\mu) \geq \mathrm{var}(\mu) \ \ (|\omega| \geq 2\,\mathrm{var}(\mu)).$$

This proves the lemma. □

Lemma 4.6.2. *Let*

$$\eta \ \mathrm{var}(\mu) < \pi/4. \tag{6.3}$$

Then all the zeros of $k(z)$ are in C_- and

$$\inf_{\omega\in\mathbb{R}} |k(i\omega)| \geq \hat{d} > 0,$$

where

$$\hat{d} := \int_0^\eta \cos(2\,\mathrm{var}(\mu)\tau) d\mu(\tau). \tag{6.4}$$

Proof. For brevity put $v = 2\,\mathrm{var}(\mu)$. Clearly,

$$\begin{aligned}|k(i\omega)|^2 &= \left| i\omega + \int_0^\eta e^{-i\omega\tau} d\mu(\tau)\right|^2 \\ &= \left(\omega - \int_0^\eta \sin(\tau\omega) d\mu(\tau)\right)^2 + \left(\int_0^\eta \cos(\tau\omega) d\mu(\tau)\right)^2.\end{aligned}$$

Hence, by the previous lemma we obtain

$$|k(i\omega)| \geq \inf_{|\omega|\leq v} |k(i\omega)| \geq \hat{d} \ \ (\omega \in \mathbb{R}).$$

Furthermore, put

$$k(m, z) = m \ \mathrm{var}(\mu)(1 + z) + (1 - m)k(z), \ 0 \leq m \leq 1.$$

We have

$$k(0, z) = k(z), k(1, z) = \mathrm{var}(\mu)(1 + z) \quad \text{and} \quad k(m, 0) = \mathrm{var}(\mu).$$

Repeating the proof of the previous lemma we derive the equality

$$\inf_{\omega\in\mathbb{R}} |k(m, i\omega)| = \inf_{-v\leq\omega\leq v} |k(m, i\omega)|.$$

In addition,

$$\mathrm{Re}\, k(m, i\omega) = \mathrm{var}(\mu)m + (1 - m) \int_0^\eta \cos(\omega\tau) d\mu.$$

Consequently,

$$|k(m, i\omega)| \geq \mathrm{var}(\mu)m + (1 - m) \int_0^\eta \cos(v\tau) d\mu.$$

Therefore,

$$|k(m, i\omega)| \geq \text{var}(\mu)m + (1-m)\hat{d} > 0 \ (\omega \in \mathbb{R}). \tag{6.5}$$

Furthermore, assume that $k(z)$ has a zero in the closed right-hand plane $\overline{C}_+$. Take into account that $k(1, z) = 1 + z$ does not have zeros in $\overline{C}_+$. So $k(m_0, i\omega)$ $(\omega \in \mathbb{R})$ should have a zero for some $m_0 \in [0, 1]$, according to continuous dependence of zeros on coefficients. But due to to (6.5) this is impossible. The proof is complete. □

Remark 4.6.3. If

$$\mu(t) - \mu(0) > 0 \text{ for some } t < \frac{\pi}{4},$$

then

$$\int_0^\eta \cos(\pi\tau)d\mu(\tau) > 0$$

and one can replace condition (6.4) by the following one:

$$\eta \ \text{var}(\mu) \leq \frac{\pi}{4}.$$

Consider the scalar function

$$k_1(z) = z + \sum_{k=1}^m b_k e^{-ih_k z} \ \ (h_k, b_k = \text{const} \geq 0).$$

The following result can be deduced from the previous lemma, but we are going to present an independent proof.

Lemma 4.6.4. *With the notation*

$$c = 2\sum_{k=1}^m b_k, \quad \textit{let} \quad h_j c < \pi/2 \ \ (j = 1, \dots, m).$$

Then all the zeros of $k_1(z)$ are in C_- and

$$\inf_{\omega \in \mathbb{R}} |k_1(i\omega)| \geq \sum_{k=1}^m b_k \ \cos\,(ch_k) > 0.$$

Proof. We restrict ourselves to the case $m = 2$. In the general case the proof is similar. Put $h_1 = v, h_2 = h$. Introduce the function

$$f(y) = |iy + b_2 e^{-ihy} + b_1 e^{-iyv}|^2.$$

Clearly,

$$\begin{aligned} f(y) &= |iy + b_2\cos(hy) + b_1\cos(yv) - i\,(b_2\sin(hy) + b_1\sin(yv))|^2 \\ &= (b_2\cos(hy) + b_1\cos(yv))^2 + (y - b_2\sin(hy) - b_1\sin(yv))^2 \\ &= y^2 + b_2^2 + b_1^2 - 2b_2 y\ \sin(hy) - 2b_1 y\sin(yv) \\ &\quad + 2b_2 b_1\ \sin(hy)\ \sin(yv) + 2b_2 b_1\cos(yv)\cos(hy). \end{aligned}$$

So

$$f(y) = y^2 + b_2^2 + b_1^2 - 2b_2 y \ \sin(hy) - 2b_1 y \sin(yv) + 2b_2 b_1 \cos y(v-h). \quad (6.6)$$

But $f(0) = (b_2 + b_1)^2$ and

$$f(y) \geq (y - b_2 - b_1)^2 \geq (b_2 + b_1)^2 \ \ (|y| > 2(b_2 + b_1)\ ).$$

Thus, the minimum of f is attained on $[-c, c]$ with $c = 2(b_2 + b_1)$. Then thanks to (6.6) $f(y) \geq w(y)$ $(0 \leq y \leq c)$, where

$$w(y) = y^2 + b_2^2 + b_1^2 - 2b_2 y \ \sin(hc) - 2b_1 y \sin(vc) + 2b_2 b_1 \cos c(h-v).$$

and

$$dw(y)/dy = 2y - 2\ (b_2 \ \sin(hc) + 2b_1 \sin(vc)).$$

The root of $dw(y)/dy = 0$ is $s = b_2 \ \sin(hc) + b_1 \sin(vc)$. Thus

$$\begin{aligned}\min_y w(y) &= s^2 + b_2^2 + b_1^2 - 2s^2 + 2b_2 b_1 \cos c(h-v) \\ &= b_2^2 + b_1^2 - (b_2 \ \sin(hc) + b_1 \sin(vc))^2 + 2b_2 b_1 \cos c(h-v).\end{aligned}$$

Hence

$$\begin{aligned}\min_y w(y) &= b_2^2 + b_1^2 - b_2^2 \sin^2(ch) - b_1^2 \sin^2(cv) + 2b_2 b_1 \cos(ch)\cos(cv) \\ &= b_2^2 \cos^2(hc) + b_1^2 \cos^2(vc) + 2b_2 b_1 \cos(ch)\cos(cv) \\ &= (b_2 \cos(ch) + b_1 \cos(cv))^2.\end{aligned}$$

This proves the required inequality. To prove stability, consider the function

$$K(z,s) = z + b_1 e^{-szv} + b_2 e^{-szh} \ \ (s \in [0,1]).$$

Then all the zeros of $K(z,0)$ are in C_- due to the just proved inequality,

$$\begin{aligned}\inf_{\omega \in \mathbb{R}} |K(i\omega, s)| &\geq b_1 \cos(csv) + b_2 \cos(csh) \\ &\geq b_1 \cos(cv) + b_2 \cos(ch) > 0 \ \ (s \in [0,1]).\end{aligned}$$

So $K(z,s)$ does not have zeros on the imaginary axis. This proves the lemma. □

Again consider equation (6.2) whose fundamental solution is defined by

$$\zeta(t) = \frac{1}{2\pi i}\int_{a-i\infty}^{a+i\infty} e^{zt}\frac{dz}{k(z)} \quad (a = \text{const}). \quad (6.7)$$

Hence,

$$\frac{1}{k(z)} = \int_0^\infty e^{-zt}\zeta(t)dt.$$

Let

$$e\eta \ \operatorname{var}(\mu) < 1. \tag{6.8}$$

Then it is not hard to show that (6.2) is exponentially stable and $\zeta(t) \geq 0$ $(t \geq 0)$ (see Section 11.4). Hence it easily follows that

$$\frac{1}{|k(i\omega)|} \leq \int_0^\infty \zeta(t)dt = \frac{1}{k(0)} \ (\omega \in \mathbb{R}).$$

But $k(0) = \operatorname{var}(\mu)$. We thus have proved the following lemma.

Lemma 4.6.5. *Let $\mu(s)$ be a nondecreasing function satisfying condition* (6.8). *Then*

$$\inf_{-\infty\leq\omega\leq\infty} \left| i\omega + \int_0^\eta \exp(-i\omega s)d\mu(s) \right| = k(0) = \operatorname{var}(\mu)$$

and

$$\int_0^\infty \zeta(t)dt = \frac{1}{k(0)} = \frac{1}{\operatorname{var}(\mu)}.$$

Now let $\mu_0(s)$ $(s \in [0, \eta\,])$ be another nondecreasing scalar function with the property

$$e\eta \ \operatorname{var}(\mu_0) \leq c_0 \ (1 \leq c_0 < 2) \tag{6.9}$$

and

$$k_0(z) = z + \int_0^\eta \exp(-zs)d\mu_0(s) \ (z \in \mathbb{C}). \tag{6.10}$$

Put $\mu(s) = \mu_0(s)/c_0$. Then

$$|k_0(i\omega) - k(i\omega)| \leq \int_0^\eta d(\mu_0(s) - \mu(s)) = (c_0 - 1) \ \operatorname{var}(\mu).$$

Hence, by the previous lemma,

$$\begin{aligned}|k_0(i\omega)| &\geq |k(i\omega)| - |k_0(i\omega) - k(i\omega)| \\ &\geq \operatorname{var}(\mu) - (c_0 - 1)\operatorname{var}(\mu) = (2 - c_0)\operatorname{var}(\mu) \ \ (\omega \in \mathbb{R}).\end{aligned}$$

We thus have proved

Lemma 4.6.6. *Let $\mu_0(s)$ $(s \in [0, \eta\,])$ be a nondecreasing scalar function satisfying* (6.9). *Then*

$$\inf_{\omega\in\mathbb{R}} \left| i\omega + \int_0^\eta \exp(-i\omega s)d\mu_0(s) \right| \geq \frac{2 - c_0}{c_0} \ \operatorname{var}(\mu_0).$$

Furthermore, let $\mu_2(s)$ be a nondecreasing function defined on $[0, \eta\,]$ and

$$k_2(z) = z + ae^{-hz} + \int_0^\eta \exp(-zs)d\mu_2(s) \ (z \in \mathbb{C})$$

with constants $a > 0$ and $h \geq 0$. Put

$$v_2 := 2(a + \mathrm{var}(\mu_2)).$$

By Lemma 4.6.1,

$$\inf_{-v_2 \leq \omega \leq v_2} |k_2(i\omega)| = \inf_{\omega \in \mathbb{R}} |k_2(i\omega)|.$$

But

$$|k_2(i\omega)|^2 = \left(\omega - a\sin(h\omega) - \int_0^\eta \sin(\tau\omega) d\mu_2(\tau)\right)^2 + \left(a\ \cos(h\omega) + \int_0^\eta \cos(\tau\omega) d\mu_2(\tau)\right)^2.$$

Let

$$v_2 h < \pi/2 \tag{6.11}$$

and

$$d_2 := \inf_{|\omega| \leq v_2} a\ \cos(h\omega) + \int_0^\eta \cos(\tau\omega) d\mu_2(\tau) > 0. \tag{6.12}$$

Then we obtain

$$\inf_{\omega \in \mathbb{R}} |k_2(i\omega)| \geq d_2. \tag{6.13}$$

Repeating the arguments of the proof of Lemma 4.6.2 we arrive at the following result.

Lemma 4.6.7. *Let conditions* (6.11) *and* (6.12) *hold. Then all the zeros of* $k_2(z)$ *are in* C_- *and inequality* (6.13) *is fulfilled.*

Let us point out one corollary of this lemma.

Corollary 4.6.8. *Let the conditions* (6.11) *and*

$$a\ \cos(hv_2) > \mathrm{var}(\mu_2)$$

hold. Then all the zeros of $k_2(z)$ *are in* C_- *and the inequality*

$$\inf_{\omega \in \mathbb{R}} |k_2(i\omega)| \geq a\cos(hv_2) - \mathrm{var}(\mu_2) > 0 \tag{6.14}$$

is valid.

In particular, if

$$k_2(z) = z + a + \int_0^\eta \exp(-zs) d\mu_2(s)\ (z \in \mathbb{C})$$

and

$$a > \mathrm{var}(\mu_2), \tag{6.15}$$

then condition (6.11) automatically holds and

$$\inf_{\omega \in \mathbb{R}} |k_2(i\omega)| \geq a - \mathrm{var}(\mu_2) > 0. \tag{6.16}$$

4.7 Systems with one distributed delay

In this section we illustrate our preceding results in the case of the equation

$$\dot{y}(t) + A \int_0^\eta y(t-s)d\mu = 0 \ \ (t \geq 0), \tag{7.1}$$

where $A = (a_{jk})$ is a constant Hurwitzian $n \times n$-matrix and μ is again a scalar nondecreasing function. So in the considered case we have $R_0(s) = \mu(s)A$,

$$K(z) = zI + \int_0^\eta e^{-zs} d\mu(s)A, \tag{7.2}$$

and the eigenvalues of matrix $K(z)$ with a fixed z are

$$\lambda_j(K(z)) = z + \int_0^\eta e^{-zs} d\mu(s)\lambda_j(A).$$

In addition,

$$g(B(i\omega)) = g\left(A \int_0^\eta e^{-i\omega s} d\mu(s)\right) \leq g(A)\,\mathrm{var}(\mu) \ \ (\omega \in \mathbb{R}),$$

$\mathrm{var}(R_0) = \|A\|_n \,\mathrm{var}(\mu)$, and $\mathrm{vd}\,(R_0) = \|A\|_n \,\mathrm{vd}\,(\mu)$, where

$$\mathrm{vd}(\mu) = \int_0^\eta \tau d\mu(\tau).$$

According to Corollary 4.3.2, we have the inequality $\theta(K) \leq \theta_A$, where

$$\theta_A := \sum_{k=0}^{n-1} \frac{(g(A)\,\mathrm{var}(\mu))^k}{\sqrt{k!}\,d_K^{k+1}}$$

with

$$d_K = \min_{j=1,\dots,n} \ \inf_{-2\,\mathrm{var}(\mu)\|A\|_n \leq \omega \leq 2\,\mathrm{var}(\mu)\|A\|_n} \left| \omega i + \lambda_j(A) \int_0^\eta e^{-i\omega s} d\mu(s) \right| .$$

In particular, if A is a normal matrix, then $g(A) = 0$ and $\theta_A = 1/d(K)$.

Now Lemma 4.4.4, (4.6), (4.7) and (4.12) imply our next result.

Theorem 4.7.1. *Let G be the fundamental solution of equation* (7.1) *and all the characteristic values of its characteristic function be in C_-. Then*

$$\|G\|_{L^2(0,\infty)} \leq W(A,\mu), \tag{7.3}$$

where

$$W(A,\mu) := \sqrt{2\theta_A\,[\|A\|_n \,\mathrm{var}(\mu)\theta_A + 1]}.$$

In addition,

$$\|\dot{G}\|_{L^2(0,\infty)} \leq \|A\|_n \operatorname{var}(\mu)\, W(A,\mu), \tag{7.4}$$

$$\|G\|_{C(0,\infty)} \leq W(A,\mu)\sqrt{2\|A\|_n \operatorname{var}(\mu)} \tag{7.5}$$

and

$$\|G\|_{L^1(0,\infty)} \leq W(A,\mu)\sqrt{\pi\theta_A(1+\|A\|_n \operatorname{vd}(\mu))}. \tag{7.6}$$

Clearly, d_K can be directly calculated. Moreover, by Lemma 4.6.3 we get the following result.

Lemma 4.7.2. *Let all the eigenvalues of A be real and positive:*

$$0 < \lambda_1(A) \leq \cdots \leq \lambda_n(A) \tag{7.7}$$

and let

$$\eta \operatorname{var}(\mu)\lambda_n(A) < \frac{\pi}{4}. \tag{7.8}$$

Then all the characteristic values of equation (7.1) *are in C_- and*

$$d_K \geq d(A,\mu), \tag{7.9}$$

where

$$d(A,\mu) := \int_0^\eta \cos\left(2\tau\lambda_n(A)\operatorname{var}(\mu)\right) d\mu(\tau) > 0.$$

Thus

$$\theta_A \leq \theta(A,\mu), \quad \text{where} \quad \theta(A,\mu) := \sum_{k=0}^{n-1} \frac{(g(A)\operatorname{var}(\mu))^k}{\sqrt{k!}\, d^{k+1}(A,\mu)}.$$

So we have proved the following result.

Corollary 4.7.3. *Let conditions* (7.7) *and* (7.8) *hold. Then all the characteristic values of equation* (7.1) *are in C_- and inequalities* (7.3)–(7.6) *are true with $\theta(A,\mu)$ instead of θ_A.*

In particular, if (7.1) takes the form

$$\dot{y}(t) + Ay(t-h) = 0 \quad (h = \text{const} > 0;\ t \geq 0), \tag{7.10}$$

then the eigenvalues of $K(z)$ are $\lambda_j(K(z)) = z + e^{-zh}\lambda_j(A)$. In addition, $\operatorname{var}(\mu) = 1$, $\operatorname{vd}(\mu) = h$. According to (7.7), condition (7.8) takes the form

$$h\lambda_n(A) < \pi/4. \tag{7.11}$$

So in this case we obtain the inequality

$$d(A,\mu) \geq \cos(2\lambda_n(A)h). \tag{7.12}$$

Furthermore, under condition (7.7), instead of condition (7.8) assume that

$$e\eta \ \mathrm{var}(\mu)\ \lambda_n(A) < 1. \tag{7.13}$$

Then according to Lemma 4.6.5 we can write

$$\inf_{\omega\in\mathbb{R}} |\lambda_j(K(i\omega))| = \inf_{\omega\in\mathbb{R}} \left| i\omega + \lambda_j(A) \int_0^\eta \exp(-i\omega s) d\mu(s) \right| = \lambda_j(A)\,\mathrm{var}(\mu).$$

Hence,

$$d_K \geq \mathrm{var}(\mu)\lambda_1(A), \tag{7.14}$$

and therefore, $\theta_A \leq \hat{\theta}_A$, where

$$\hat{\theta}_A := \frac{1}{\mathrm{var}(\mu)} \sum_{k=0}^{n-1} \frac{g^k(A)}{\sqrt{k!}\lambda_1^{k+1}(A)}.$$

Now taking into account Theorem 4.7.1 we arrive at our next result.

Corollary 4.7.4. *Let conditions* (7.7) *and* (7.13) *hold. Then all the characteristic values of equation* (7.1) *are in* C_- *and inequalities* (7.3)–(7.6) *are true with* $\theta_A = \hat{\theta}_A$.

Now let A be diagonalizable. Then there is an invertible matrix T and a normal matrix S, such that $T^{-1}AT = S$. Besides,

$$T^{-1}K(z)T = K_S(z) \quad \text{where} \quad K_S(z) = zI + \int_0^\eta e^{-zs} d\mu(s) S$$

and therefore,

$$\|K^{-1}(z)\|_n \leq \kappa_T \|K_S^{-1}(z)\|_n$$

where $\kappa_T = \|T^{-1}\|_n \|T\|_n$. Recall that some estimates for κ_T are given in Section 2.7.

Since $g(S) = 0$, and the eigenvalues of K and K_S coincide, we have $d_K = d_{K_S}$,

$$\theta_S = \frac{1}{d_K} \quad \text{and therefore} \quad \theta_A \leq \frac{\kappa_T}{d_K}.$$

Now Theorem 4.7.1 implies the following result.

Corollary 4.7.5. *Suppose A is diagonalizable and all the characteristic values of equation* (7.1) *are in* C_-. *Then inequalities* (7.3)–(7.6) *are true with* $\frac{\kappa_T}{d(K)}$ *instead of* θ_A.

If, in addition, conditions (7.7) and (7.8) hold, then according to (7.9) equation (7.1) is stable and

$$\theta_A \leq \frac{\kappa_T}{d(A,\mu)}. \tag{7.15}$$

Moreover, if conditions (7.7) and (7.13) hold, then according to (7.14) equation (7.1) is also stable and

$$\theta_A \leq \frac{\kappa_T}{\lambda_1(A)\,\mathrm{var}(\mu)}. \tag{7.16}$$

Note also that if A is Hermitian, and conditions (7.7) and (7.13) hold, then reducing (7.1) to the diagonal form, due to Lemma 4.6.5, we can assert that the fundamental solution G to (7.1) satisfies the inequality

$$\|G\|_{L^1(0,\infty)} \leq \frac{1}{\mathrm{var}(\mu)} \sum_{k=1}^{n} \frac{1}{\lambda_k(A)}. \tag{7.17}$$

So, if A is diagonalizable and conditions (7.7) and (7.13) holds, then

$$\|G\|_{L^1(0,\infty)} \leq \frac{\kappa_T}{\mathrm{var}(\mu)} \sum_{k=1}^{n} \frac{1}{\lambda_k(A)}. \tag{7.18}$$

4.8 Estimates via determinants

Again consider equation (1.1). As it was shown in Section 2.3,

$$\|A^{-1}\|_n \leq \frac{N_2^{n-1}(A)}{(n-1)^{(n-1)/2}|\det(A)|} \quad (n \geq 2),$$

for any invertible $n \times n$-matrix A. The following result directly follows from this inequality.

Corollary 4.8.1. *For any regular point z of the characteristic function of equation* (1.1), *one has*

$$\|K^{-1}(z)\|_n \leq \frac{N_2^{n-1}(K(z))}{(n-1)^{(n-1)/2}|\det(K(z))|} \quad (n \geq 2), \tag{8.1}$$

and thus $\theta(K) \leq \theta_{\det}(K)$ where

$$\theta_{\det}(K) := \sup_{-2\,\mathrm{var}(R_0) \leq \omega \leq 2\,\mathrm{var}(R_0)} \frac{N_2^{n-1}(K(i\omega))}{(n-1)^{(n-1)/2}|\det(K(i\omega))|}.$$

This corollary is more convenient for the calculations than (3.6), if n is small enough.

Recall that it is assumed that all the characteristic values of (1.1) are in the open left half-plane C_-. Let

$$W_{\det}(K) := \sqrt{2\theta_{\det}(K)[\mathrm{var}(R_0)\theta_{\det}(K) + 1]}.$$

By Lemma 4.4.4 we arrive at the following lemma.

Lemma 4.8.2. *The inequality* $\|G\|_{L^2(0,\infty)} \le W_{\text{det}}(K)$ *is valid.*

In addition, Lemma 4.4.5 and Theorem 4.4.7 imply our next result.

Lemma 4.8.3. *The inequalities*

$$\|\dot{G}\|_{L^2(0,\infty)} \le W_{\text{det}}(K)\,\text{var}(R_0)$$

and

$$\|G\|_{C(0,\infty)} \le a_{\text{det}}(K), \tag{8.2}$$

hold, where

$$a_{\text{det}}(K) := \sqrt{2\,\text{var}(R_0)}W_{\text{det}}(K).$$

Clearly,

$$a_{\text{det}}(K) = 2\sqrt{\text{var}(R_0)\theta_{\text{det}}(K)[\text{var}(R_0)\theta_{\text{det}}(K)+1]}$$

and

$$a_{\text{det}}(K) \le 2(1+\text{var}(R_0)\theta_{\text{det}}(K)) \le 2(1+\text{var}(R_0)\hat{\theta}_{\text{det}}(K)).$$

To estimate the L^1-norm of the fundamental solution via the determinant, we use inequality (4.12) and Corollary 4.8.1, by which we get the following result.

Corollary 4.8.4. *The fundamental solution* G *to equation* (1.1) *satisfies the inequality*

$$\|G\|_{L^1(0,\infty)} \le W_{\text{det}}(K)\sqrt{\pi\theta_{\text{det}}(K)(1+\text{vd}\,(R_0))}.$$

Furthermore, if the entries $r_{jk}(\tau)$ of $R_0(\tau)$ have the following properties: r_{jj} are non-increasing and $r_{jk}(\tau) \equiv 0$ $(j > k)$, that is, R_0 is triangular; then clearly,

$$\det\ K(z) = \prod_{k=1}^{n}\left(z - \int_0^\eta e^{-zs}dr_{kk}(s)\right). \tag{8.3}$$

If, in addition

$$\eta\ \text{var}\, r_{jj} \le \pi/4, j = 1, \dots, n, \tag{8.4}$$

then by Lemma 4.6.2 all the zeros of det $K(z)$ are in C_- and

$$|\det\ K(i\omega)| \ge \prod_{k=1}^{n} \hat{d}_{kk} > 0 \tag{8.5}$$

where

$$\hat{d}_{jj} = \int_0^\eta \cos(2\,\text{var}(r_{jj})\tau)\,dr_{jj}(\tau).$$

4.9 Stability of diagonally dominant systems

Again $r_{jk}(s)$ $(j,k=1,\dots,n)$ are the entries of $R_0(s)$. In addition $r_{jj}(s)$ are non-increasing. Put

$$\xi_j = \sum_{k=1,k\neq j}^{n} \operatorname{var}(r_{jk}).$$

Lemma 4.9.1. *Assume that all the zeros of the functions*

$$k_j(z) := z - \int_0^\eta e^{-zs} dr_{jj}(s),\ j=1,\dots,n,$$

are in C_- and, in addition,

$$|k_j(i\omega)| > \xi_j \quad (\omega \in \mathbb{R};\ |\omega| \le 2\operatorname{var}(r_{jj});\ j=1,\dots,n), \tag{9.1}$$

then equation (1.1) *is exponentially stable. Moreover,*

$$|\det(K(i\omega))| \ge \prod_{j=1}^{n} (|k_j(i\omega)| - \xi_j) \quad (\omega \in \mathbb{R}).$$

Proof. Clearly

$$\sum_{k=1,k\neq j}^{n} \left| \int_0^\eta e^{-i\omega s} dr_{jk}(s) \right| \le \xi_j.$$

Hence the result is due to the Ostrowski theorem (see Section 1.11). □

If condition (8.4) holds, then by Lemma 4.6.2 we obtain

$$|k_j(i\omega)| \ge \hat{d}_{jj} > 0$$

where $\hat{d}_{jj}$ are defined in the previous section. So due to the previous lemma we arrive at the following result.

Corollary 4.9.2. *If the conditions* (8.4) *and*

$$\hat{d}_{jj} > \xi_j \quad (j=1,\dots,n)$$

hold, then equation (1.1) *is exponentially stable. Moreover,*

$$|\det(K(i\omega))| \ge \prod_{j=1}^{n} (\hat{d}_{jj} - \xi_j) \quad (\omega \in \mathbb{R}).$$

Now we can apply the results of the previous section.

4.10 Comments

The contents of this chapter is based on the paper [26] and on some results from Chapter 8 of the book [24].

Chapter 5

Properties of Characteristic Values

In this chapter we investigate properties of the characteristic values of autonomous systems. In particular some identities for characteristic values and perturbations results are derived.

5.1 Sums of moduli of characteristic values

Recall that

$$K(z) = zI - \int_0^{\eta} \exp(-zs) dR_0(s)$$

is the characteristic matrix function of the autonomous equation

$$\dot{y} = \int_0^{\eta} dR_0(s) y(t-s).$$

Here and below in this chapter $R_0(\tau)$ is an $n \times n$-matrix-valued function defined on a finite segment $[0, \eta]$, whose entries have bounded variations and I is the unit matrix.

Enumerate the characteristic values $z_k(K)$ of K with their multiplicities in the nondecreasing order of their absolute values: $|z_k(K)| \le |z_{k+1}(K)|$ $(k = 1, 2, \dots)$. If K has $l < \infty$ zeros, we put $1/z_k(K) = 0$ $(k = l+1, l+2, \dots)$.

The aim of this section is to estimate the sums

$$\sum_{k=1}^{j} \frac{1}{|z_k(K)|} \quad (j = 1, 2, \dots).$$

To this end, note that

$$K(z) = zI - \frac{(-1)^k z^k}{k!} \int_0^{\eta} s^k dR_0(s) = \sum_{k=0}^{\infty} \frac{z^k}{k!} B_k, \tag{1.1}$$

where

$$B_0 = -\int_0^{\eta} dR_0(s) = -R_0(\eta\,), \quad B_1 = I + \int_0^{\eta} s dR_0(s),$$
$$B_k = (-1)^{k+1} \int_0^{\eta} s^k dR_0(s) \ \ (k \geq 2).$$

In addition, it is supposed that

$$R_0(\eta\,) \text{ is invertible.} \tag{1.2}$$

Put $C_k = -R_0^{-1}(\eta\,)B_k$ and

$$K_1(z) := -R_0^{-1}(\eta\,)K(z) = \sum_{k=0}^{\infty} \frac{z^k}{k!} C_k \ \ (C_0 = I).$$

Since det $K_1(z) = -\det\ R_0^{-1}(\eta\,)\ \det\ K(z)$, all the characteristic values of K and K_1 coincide.

For brevity, in this chapter $\|A\|$ denotes the spectral norm of a matrix A: $\|A\| = \|A\|_n$. Without loss of generality assume that

$$\eta\, < 1. \tag{1.3}$$

If this condition is not valid, by the substitution $z = wa$ with some $a > \eta$ into (1.1), we obtain

$$K(aw) = awI - \int_0^{\eta} \exp(-saw)dR_0(s)$$
$$= a\left(wI - \frac{1}{a}\int_0^{a\eta} \exp(-\tau w)dR_0(\tau/a)\right) = aK_1(w),$$

where

$$K_1(w) = wI - \int_0^{\eta_{\,1}} \exp(-\tau w)dR_1(\tau)$$

with $R_1(\tau) = \frac{1}{a}R_0(\tau/a)$ and $\eta_{\,1} = \frac{\eta}{a}$. Thus, condition (1.3) holds with $\eta\, = \eta_{\,1}$.

Let $r_{jk}(s)$ be the entries of $R_0(s)$. Recall that var(R_0) denotes the spectral norm of the matrix $(\text{var}(r_{jk}))_{j,k=1}^n$:

$$\text{var}(R_0) = \|\ (\text{var}(r_{jk}))_{j,k=1}^n\ \|.$$

Lemma 5.1.1. *The inequality*

$$\|B_k\| = \left\|\int_0^{\eta} s^k dR_0(s)\right\| \leq \eta^{\,k}\, \text{var}(R_0) \ \ (k \geq 2)$$

is true.

The proof of this lemma is similar to the proof of Lemma 4.4.8. It is left to the reader.

From this lemma it follows that

$$\|C_k\| \leq \mathrm{var}(R_0)\|R_0^{-1}(\eta\,)\|\eta^{\,k} \quad (k \geq 2).$$

So under condition (1.3) we have

$$\sum_{k=1}^{\infty} \|C_k\|^2 < \infty. \tag{1.4}$$

Furthermore, let

$$\Psi_K := \left[\sum_{k=1}^{\infty} C_k C_k^*\right]^{1/2}.$$

So Ψ_K is an $n \times n$-matrix. Set

$$\omega_k(K) = \begin{cases} \lambda_k(\Psi_K) & \text{for } k = 1, \ldots, n, \\ 0 & \text{if } k \geq n+1, \end{cases}$$

where $\lambda_k(\Psi_K)$ are the eigenvalues of matrix Ψ_K with their multiplicities and enumerated in the decreasing way: $\omega_k(K) \geq \omega_{k+1}(K)$ $(k = 1, 2, \ldots)$.

Theorem 5.1.2. *Let conditions* (1.2) *and* (1.3) *hold. Then the characteristic values of K satisfy the inequalities*

$$\sum_{k=1}^{j} \frac{1}{|z_k(K)|} < \sum_{k=1}^{j} \left[\omega_k(K) + \frac{n}{k+n}\right] \quad (j = 1, 2, \ldots).$$

This result is a particular case of Theorem 12.2.1 [46].

From the latter theorem it follows that

$$\frac{j}{|z_j(K)|} < \sum_{k=1}^{j} \left[\omega_k(K) + \frac{n}{k+n}\right]$$

and thus,

$$|z_j(K)| > \frac{j}{\sum_{k=1}^{j} \left[\omega_k(K) + \frac{n}{(k+n)}\right]} \quad (j = 1, 2, \ldots).$$

Therefore, in the disc

$$|z| \leq \frac{j}{\sum_{k=1}^{j} \left[\omega_k(K) + \frac{n}{(k+n)}\right]} \quad (j = 2, 3, \ldots)$$

K has no more than $j-1$ characteristic values. Let $\nu_K(r)$ be the function counting the characteristic values of K in the disc $|z| \leq r$.

We consequently get

Corollary 5.1.3. *Let conditions* (1.2) *and* (1.3) *hold. Then the inequality* $\nu_K(r) \le j-1$ *is valid, provided*

$$r \le \frac{j}{\sum_{k=1}^{j}\left[\omega_k(K) + \frac{n}{(k+n)}\right]} \quad (j = 2, 3, \dots).$$

Moreover, $K(z)$ *does not have characteristic values in the disc*

$$|z| \le \frac{1}{\lambda_1(\Psi_K) + \frac{n}{1+n}}.$$

Let us apply the Borel (Laplace) transform. Namely, put

$$F_K(w) = \int_0^\infty e^{-wt} K_1(t)dt \ \ (w \in \mathbb{C};\ |w| = 1).$$

Then

$$F_K(w) = -R_0^{-1}(\eta)\int_0^\infty e^{-wt}\left(tI - \int_0^\eta \exp(-ts)dR_0(s)\right)dt.$$

We can write down

$$F_K(w) = -R_0^{-1}(\eta)\left[\frac{1}{w^2}I - \int_0^\eta \frac{1}{s+w}dR_0(s)\right]. \tag{1.5}$$

On the other hand

$$F_K(w) = \sum_{k=0}^\infty \frac{1}{w^{k+1}}C_k$$

and therefore

$$\frac{1}{2\pi}\int_0^{2\pi} F_K(e^{-is})F_K^*(e^{is})ds = \sum_{k=0}^\infty C_k C_k^*.$$

Thus, we have proved the following result.

Lemma 5.1.4. *Let condition* (1.3) *hold. Then*

$$\Psi_K^2 = \frac{1}{2\pi}\int_0^{2\pi} F_K(e^{-is})F_K^*(e^{is})ds$$

and consequently

$$\|\Psi_K\|^2 \le \frac{1}{2\pi}\int_0^{2\pi} \|F_K(e^{is})\|^2 ds \le \sup_{|w|=1}\|F_K(w)\|^2.$$

Simple calculations show that

$$\left\| \int_0^{\eta} \frac{1}{s+w} dR_0(s) \right\| \leq \frac{\mathrm{var}(R_0)}{1-\eta} \quad (|w|=1).$$

Taking into account (1.5), we deduce that

$$\|\Psi_K\| \leq \|R_0^{-1}\| \left(1 + \frac{\mathrm{var}(R_0)}{1-\eta}\right)$$

and consequently,

$$\omega_j(K) \leq \sup_{|w|=1} \|F_K(w)\| \leq \alpha(F_K), \tag{1.6}$$

where

$$\alpha(F_K) := \|R_0^{-1}\| \left(1 + \frac{\mathrm{var}(R_0)}{1-\eta}\right) \quad (j \leq n).$$

Note that the norm of $R_0^{-1}(\eta)$ can be estimated by the results presented in Section 2.3 above. Inequality (1.6) and Theorem 5.1.2 imply the following result.

Corollary 5.1.5. *Let conditions* (1.2) *and* (1.3) *hold. Then the characteristic values of K satisfy the inequalities*

$$\sum_{k=1}^{j} \frac{1}{|z_k(K)|} < j\alpha(F_K) + n \sum_{k=1}^{j} \frac{1}{k+n} \quad (j \leq n)$$

and

$$\sum_{k=1}^{j} \frac{1}{|z_k(K)|} < n \left(\alpha(F_K) + \sum_{k=1}^{j} \frac{1}{k+n} \right) \quad (j > n).$$

5.2 Identities for characteristic values

This section deals with the following sums containing the characteristic values $z_k(K)$ $(k = 1, 2, \dots)$:

$$\tilde{s}_m(K) := \sum_{k=1}^{\infty} \frac{1}{z_k^m(K)} \quad (m = 2, 3, \dots),$$

$$\sum_{k=1}^{\infty} \left(\mathrm{Im} \frac{1}{z_k(K)} \right)^2 \quad \text{and} \quad \sum_{k=1}^{\infty} \left(\mathrm{Re} \frac{1}{z_k(K)} \right)^2.$$

Again use the expansion

$$K_1(z) := -R_0^{-1}(\eta) K(z) = \sum_{k=0}^{\infty} \frac{z^k}{k!} C_k, \tag{2.1}$$

where

$$C_k = (-1)^k R_0^{-1}(\eta) \int_0^\eta s^k dR_0(s) \ (k \geq 2),$$

$$C_0 = I \quad \text{and} \quad C_1 = -R_0^{-1}(\eta) \left(I + \int_0^\eta s dR_0(s) \right).$$

To formulate our next result, for an integer $m \geq 2$, introduce the $m \times m$-block matrix

$$\hat{B}_m = \begin{pmatrix} -C_1 & -C_2/2 & \dots & -C_{m-1}/(m-1)! & -C_m/m! \\ I & 0 & \dots & 0 & 0 \\ 0 & I & \dots & 0 & 0 \\ \cdot & \cdot & \dots & \cdot & \cdot \\ 0 & 0 & \dots & I & 0 \end{pmatrix}$$

whose entries are $n \times n$ matrices.

Theorem 5.2.1. *For any integer $m \geq 2$, we have*

$$\tilde{s}_m(K) = \operatorname{Trace} \hat{B}_m^m.$$

This result is a particular case of Theorem 12.7.1 from [46].
Furthermore, let

$$\tau(K) := \sum_{k=1}^{\infty} N_2^2(C_k) + (\zeta(2) - 1)n$$

and

$$\psi(K,t) := \tau(K) + \operatorname{Re} \operatorname{Trace}[e^{2it} \ (C_1^2 - C_2/2)] \ \ (t \in [0, 2\pi))$$

where

$$\zeta(z) := \sum_{k=1}^{\infty} \frac{1}{k^z} \ \ (\operatorname{Re} z > 1)$$

is the Riemann zeta function.

Theorem 5.2.2. *Let conditions* (1.2) *and* (1.3) *hold. Then for any $t \in [0, 2\pi)$ the relations*

$$\tau(K) - \sum_{k=1}^{\infty} \frac{1}{|z_k(K)|^2} = \psi(K,t) - 2 \sum_{k=1}^{\infty} \left(\operatorname{Re} \frac{e^{it}}{z_k(K)} \right)^2 \geq 0$$

are valid.

This theorem is a particular case of Theorem 12.4.1 from [46].
Note that

$$\psi(K, \pi/2) = \tau(K) - \operatorname{Re} \operatorname{Trace} \left(C_1^2 - \frac{1}{2} C_2 \right)$$

and

$$\psi(K,0) = \tau(K) + \mathrm{Re\,Trace}\left(C_1^2 - \frac{1}{2}C_2\right).$$

Now Theorem 5.2.2 yields the following result.

Corollary 5.2.3. *Let conditions* (1.2) *and* (1.3) *hold. Then*

$$\tau(K) - \sum_{k=1}^{\infty} \frac{1}{|z_k(K)|^2} = \psi(K,\pi/2) - 2\sum_{k=1}^{\infty}\left(\mathrm{Im}\,\frac{1}{z_k(K)}\right)^2$$
$$= \psi(K,0) - 2\sum_{k=1}^{\infty}\left(\mathrm{Re}\,\frac{1}{z_k(K)}\right)^2 \geq 0.$$

Consequently,

$$\sum_{k=1}^{\infty} \frac{1}{|z_k(K)|^2} \leq \tau(K), \quad 2\sum_{k=1}^{\infty}\left(\mathrm{Im}\,\frac{1}{z_k(K)}\right)^2 \leq \psi(f,\pi/2)$$

and

$$2\sum_{k=1}^{\infty}\left(\mathrm{Re}\,\frac{1}{z_k(K)}\right)^2 \leq \psi(f,0).$$

5.3 Multiplicative representations of characteristic functions

Let A be an $n \times n$-matrix and E_k $(k = 1,\dots,n)$ be the maximal chain of the invariant orthogonal projections of A. That is,

$$0 = E_0 \subset E_1 \subset \cdots \subset E_n = I$$

and

$$AE_k = E_k A E_k \ (k = 1,\dots,n). \tag{3.1}$$

Besides, $\Delta E_k = E_k - E_{k-1}$ $(k = 1,\dots,n)$ are one-dimensional. Again set

$$\prod_{1\leq k\leq m}^{\rightarrow} X_k := X_1 X_2 \dots X_m$$

for matrices $X_1, X_2, \dots, X_m$. As it is shown in Subsection 2.2.2,

$$(I - A)^{-1} = \prod_{1\leq k\leq n}^{\rightarrow}\left(I + \frac{A\Delta E_k}{1 - \lambda_k(A)}\right), \tag{3.2}$$

provided $I - A$ is invertible.

Furthermore, for each fixed $z \in \mathbb{C}$, $K(z)$ possesses the maximal chain of invariant orthogonal projections, which we denote by $E_k(K,z)$:

$$0 = E_0(K,z) \subset E_1(K,z) \subset \cdots \subset E_n(K,z) = I$$

and

$$K(z)E_k(K,z) = E_k(K,z)K(z)E_k(K,z) \ (k = 1,\dots,n).$$

Moreover,

$$\Delta E_k(K,z) = E_k(K,z) - E_{k-1}(K,z) \ \ (k = 1,\dots,n)$$

are one-dimensional orthogonal projections.

Write $K(z) = zI - B(z)$ with

$$B(z) = \int_0^\eta \exp(-zs)dR_0(s).$$

Then $K(z) = z(I - \frac{1}{z}B(z))$. Now by (3.2) with $K(z)$ instead of $I - A$, we get our next result.

Theorem 5.3.1. *For any regular $z \neq 0$ of K, the equality*

$$K^{-1}(z) = \frac{1}{z} \prod_{1 \le k \le n}^{\rightarrow} \left(I + \frac{B(z)\Delta E_k(z)}{z\lambda_k(K(z))} \right)$$

is true.

Furthermore, let

$$A = D + V \ \ (\sigma(A) = \sigma(D)) \tag{3.3}$$

be the Schur triangular representation of A (see Section 2.2). Namely, V is the nilpotent part of A and

$$D = \sum_{k=1}^{n} \lambda_k(A)\Delta E_k$$

is its diagonal part. Besides $VE_k = E_kVE_k$ $(k = 1,\dots,n)$. Let us use the equality

$$A^{-1} = D^{-1} \prod_{2 \le k \le n}^{\rightarrow} \left[I - \frac{V\Delta E_k}{\lambda_k(A)} \right]$$

for any non-singular matrix A (see Section 2.2). Now replace A by $K(z)$ and denote by $\tilde{D}_K(z)$ and $\tilde{V}_K(z)$ be the diagonal and nilpotent parts of $K(z)$, respectively. Then the previous equality at once yields the following result.

Theorem 5.3.2. *For any regular z of K, the equality*

$$K^{-1}(z) = \tilde{D}_K^{-1}(z) \prod_{2 \le k \le n}^{\rightarrow} \left[I - \frac{\tilde{V}_K(z)\Delta E_k(z)}{\lambda_k(K(z))} \right]$$

is true.

5.4 Perturbations of characteristic values

Let

$$K(z) = zI - \int_0^{\eta} \exp(-zs)dR_0(s) \tag{4.1a}$$

and

$$\tilde{K}(z) = zI - \int_0^{\eta} \exp(-zs)d\tilde{R}(s), \tag{4.1b}$$

where R_0 and $\tilde{R}$ are $n \times n$-matrix-valued functions defined on $[0, \eta]$, whose entries have bounded variations.

Enumerate the characteristic values $z_k(K)$ and $z_k(\tilde{K})$ of K and $\tilde{K}$, respectively with their multiplicities in the nondecreasing order of their moduli.

The aim of this section is to estimate the quantity

$$rv_K(\tilde{K}) = \max_{j=1,2,\dots} \min_{k=1,2,\dots} \left| \frac{1}{z_k(K)} - \frac{1}{z_j(\tilde{K})} \right|$$

which will be called *the relative variation of the characteristic values of $\tilde{K}$ with respect to the characteristic values of K.*

Assume that

$$R_0(\eta) \quad \text{and} \quad \tilde{R}(\eta) \text{ are invertible,} \tag{4.2}$$

and

$$\eta < 1. \tag{4.3}$$

As it was shown in Section 5.1, the latter condition does not affect the generality. Again put

$$K_1(z) = -R^{-1}(\eta)K(z) \quad \text{and, in addition,} \quad \tilde{K}_1(z) = -\tilde{R}^{-1}(\eta)\tilde{K}(z).$$

So

$$K_1(z) = \sum_{k=0}^{\infty} \frac{z^k}{k!} C_k \quad \text{and} \quad \tilde{K}_1(z) = \sum_{k=0}^{\infty} \frac{z^k}{k!} \tilde{C}_k, \tag{4.4}$$

where

$$C_k = (-1)^k R_0^{-1}(\eta) \int_0^{\eta} s^k dR_0(s)$$

and

$$\tilde{C}_k = (-1)^k \tilde{R}^{-1}(\eta) \int_0^{\eta} s^k d\tilde{R}(s) \ (k \geq 2);$$

$C_0 = \tilde{C}_0 = I$;

$$C_1 = -R_0^{-1}(\eta) \left(I + \int_0^{\eta} s dR_0(s) \right) \quad \text{and} \quad \tilde{C}_1 = -\tilde{R}^{-1}(\eta) \left(I + \int_0^{\eta} s d\tilde{R}(s) \right).$$

As it is shown in Section 5.1, condition (4.3) implies the inequalities

$$\sum_{k=0}^{\infty} \|C_k\|^2 < \infty \quad \text{and} \quad \sum_{k=0}^{\infty} \|\tilde{C}_k\|^2 < \infty. \tag{4.5}$$

Recall that

$$\Psi_K := \left[\sum_{k=1}^{\infty} C_k C_k^*\right]^{1/2}$$

and put

$$w(K) := 2N_2(\Psi_K) + 2[n(\zeta(2) - 1)]^{1/2},$$

where $\zeta(.)$ is again the Riemann zeta function. Also let

$$\xi(K,s) := \frac{1}{s} \exp\left[\frac{1}{2} + \frac{w^2(K)}{2s^2}\right] \quad (s > 0) \tag{4.6}$$

and

$$q = \left[\sum_{k=1}^{\infty} \|\tilde{C}_k - C_k\|^2\right]^{1/2}.$$

Theorem 5.4.1. *Let conditions* (4.2) *and* (4.3) *hold. Then*

$$rv_K(\tilde{K}) \le r(K,q),$$

where $r(K,q)$ *is the unique positive root of the equation*

$$q\xi(K,s) = 1. \tag{4.7}$$

This result is a particular case of Theorem 12.5.1 from [46]. If we substitute the equality $y = xw(K)$ into (4.7) and apply Lemma 1.6.4 from [46], we get $r(K,q) \le \delta(K,q)$, where

$$\delta(K,q) := \begin{cases} e\,q & \text{if } w(K) \le e\,q, \\ w(K)\,[ln\,(w(K)/q)]^{-1/2}. & \text{if } w(K) > e\,q \end{cases}$$

Therefore, $rv_K(\tilde{K}) \le \delta(K,q)$.

Put

$$W_j = \left\{z \in \mathbb{C} : q\xi\left(K, \left|\frac{1}{z} - \frac{1}{z_j(K)}\right|\right) \ge 1\right\} \quad (j = 1, 2, \ldots).$$

Since $\xi(K,y)$ is a monotone decreasing function with respect to $y > 0$, Theorem 5.4.1 yields the following result.

Corollary 5.4.2. *Under the hypothesis of Theorem* 5.4.1, *all the characteristic values of* $\tilde{K}$ *lie in the set*

$$\cup_{j=1}^{\infty} W_j.$$

Let us apply the Borel (Laplace) transform. Namely, put

$$F(u) = \int_0^\infty e^{-ut}(\tilde{K}_1(t) - K_1(t))dt \quad (u \in \mathbb{C};\ |u| = 1).$$

Then

$$F(u) = \sum_{k=0}^{\infty} \frac{1}{u^{k+1}}(\tilde{C}_k - C_k).$$

Therefore

$$\frac{1}{2\pi}\int_0^{2\pi} F^*(e^{-is})F(e^{is})ds = \sum_{k=0}^{\infty}((\tilde{C}_k)^* - C_k^*)(\tilde{C}_k - C_k).$$

Thus,

$$\text{Trace}\,\frac{1}{2\pi}\int_0^{2\pi} F^*(e^{-is})F(e^{is})ds = \text{Trace}\sum_{k=0}^{\infty}((\tilde{C})_k^* - C_k^*)(\tilde{C}_k - C_k),$$

or

$$q^2 \le \sum_{k=0}^{\infty} N_2^2(\tilde{C}_k - C_k) = \frac{1}{2\pi}\int_0^{2\pi} N_2^2(F(e^{is}))ds.$$

This forces

$$q^2 \le \frac{1}{2\pi}\int_0^{2\pi} N_2^2(F(e^{is}))ds \le \sup_{|z|=1} N_2^2(F(z)).$$

5.5 Perturbations of characteristic determinants

In the present section we investigate perturbations of characteristic determinants. Besides, K and $\tilde{K}$ are defined by (4.1), again.

Lemma 5.5.1. *The equality*

$$\inf_{-\infty\le\omega\le\infty} |\det\ K(i\omega)| = \inf_{-2\ \text{var}(R_0)\le\omega\le 2\ \text{var}(R_0)} |\det\ K(i\omega)|$$

is valid. Moreover,

$$\inf_{-\infty\le\omega\le\infty} |\det\ K(i\omega)| \le |\det\ R_0(\eta\,)|.$$

Proof. Put $d(s) = |\det\ K(is)|$. Clearly, $d(0) = |\det\ R_0(\eta)|$ and

$$\det\ K(z) = \prod_{k=1}^{n} \lambda_k(zI - B(z)) = \prod_{k=1}^{n} (z - \lambda_k(B(z)).$$

In addition,

$$|\lambda_k(B(is))| \le \|B(is)\| \le \operatorname{var}(R_0)$$

and thus

$$d(s) \ge ||s| - \operatorname{var}(R_0)|^n \ge (\operatorname{var}(R_0))^n \quad (|s| \ge 2\operatorname{var}(R_0)).$$

We conclude that the minimum of $d(s)$ is attained inside the segment

$$[-2\ \operatorname{var}(R_0), 2\ \operatorname{var}(R_0)],$$

as claimed. □

Furthermore, let A and B be $n \times n$-matrices. As it is shown in Section 1.11,

$$|\det\ A - \det\ B| \le \|A - B\| \left[1 + \frac{1}{2}\|A - B\| + \frac{1}{2}\|A + B\|\right]^n.$$

Hence we get

Corollary 5.5.2. *The inequality*

$$|\det K(z) - \det \tilde{K}(z)| \le \|K(z) - \tilde{K}(z)\| \left[1 + \frac{1}{2}\|K(z) - \tilde{K}(z)\| + \frac{1}{2}\|K(z) + \tilde{K}(z)\|\right]^n$$

holds.

Simple calculations show that

$$\|K(i\omega)\| \le |\omega| + \operatorname{var}(R_0), \ \|K(i\omega) + \tilde{K}(i\omega)\| \le 2|\omega| + \operatorname{var}(R_0) + \operatorname{var}(\tilde{R})$$

and

$$\|K(i\omega) - \tilde{K}(i\omega)\| \le \operatorname{var}(R_0 - \tilde{R}) \ \ (\omega \in \mathbb{R}).$$

We thus arrive at the following corollary.

Corollary 5.5.3. *The inequality*

$$|\det K(i\omega) - \det \tilde{K}(i\omega)| \le \operatorname{var}(R_0 - \tilde{R}) \left[1 + |\omega| + \frac{1}{2}(\operatorname{var}(R_0 + \tilde{R}) + \operatorname{var}(R_0 - \tilde{R}))\right]^n$$

is valid.

In particular,

$$|\det\ \tilde{K}(i\omega)| \ge |\det\ K(i\omega)| - \operatorname{var}(R_0 - \tilde{R}) \left[1 + |\omega| + \frac{1}{2}(\operatorname{var}(R_0 + \tilde{R}) + \operatorname{var}(R_0 - \tilde{R}))\right]^n.$$

Now simple calculations and Lemma 5.5.1 imply the following result.

Theorem 5.5.4. *Let all the zeros of* det $K(z)$ *be in* C_- *and*

$$J_1 := \inf_{|\omega|\le 2\,\mathrm{var}(R_0)} |\det\ K(i\omega)| - \mathrm{var}(R_0 - \tilde{R})\left[1 + 2\,\mathrm{var}(\tilde{R}) + \frac{1}{2}\left(\mathrm{var}(R_0 + \tilde{R}) + \mathrm{var}(R_0 - \tilde{R})\right)\right]^n > 0.$$

Then all the zeros of det $\tilde{K}(z)$ *are also in* C_- *and*

$$\inf_{\omega\in\mathbb{R}} |\det\ \tilde{K}(i\omega)| \ge J_1.$$

Note that

$$J_1 \ge \inf_{|\omega|\le 2\,\mathrm{var}(R_0)} |\det\ K(i\omega)| - \mathrm{var}(R_0 - \tilde{R})[1 + 3\,\mathrm{var}(\tilde{R}) + \mathrm{var}(\tilde{R})]^n.$$

Furthermore, let us establish a result similar to Theorem 5.5.1 but in the terms of the norm $N_2(.)$. Again let A and B be $n \times n$-matrices. Then as it is shown in Section 1.11,

$$|\det\ A - \det\ B| \le \frac{1}{n^{n/2}} N_2(A - B)\left[1 + \frac{1}{2}N_2(A - B) + \frac{1}{2}N_2(A + B)\right]^n.$$

Now at once we get our next result.

Lemma 5.5.5. *The inequality*

$$|\det\ K(z) - \det\ \tilde{K}(z)| \le \frac{1}{n^{n/2}} N_2(K(z) - \tilde{K}(z))\left[1 + \frac{1}{2}N_2(K(z) - \tilde{K}(z)) + \frac{1}{2}N_2(K(z) + \tilde{K}(z))\right]^n$$

is valid.

Note that the previous lemma implies the inequality

$$|\det\ K(z) - \det\ \tilde{K}(z)| \le \frac{1}{n^{n/2}} N_2(K(z) - \tilde{K}(z))\left[1 + N_2(K(z)) + N_2(\tilde{K}(z))\right]^n.$$

To illustrate the results obtained in this section consider the following matrix-valued functions:

$$K(z) = zI - \int_0^\eta e^{-sz} A(s)ds - \sum_{k=0}^m e^{-h_k z} A_k, \tag{5.2a}$$

and

$$\tilde{K}(z) = zI - \int_0^\eta e^{-sz} \tilde{A}(s)ds - \sum_{k=0}^m e^{-h_k z} \tilde{A}_k, \tag{5.2b}$$

where $A(s)$ and $\tilde{A}(s)$ are integrable matrix functions, A_k and $\tilde{A}_k$ are constant matrices, h_k are positive constants. Put

$$\mathrm{Var}_2(R_0) = \int_0^\eta N_2(A(s))ds + \sum_{k=0}^{m} N_2(A_k),$$

$$\mathrm{Var}_2(R_0 \pm \tilde{R}) = \int_0^\eta N_2(A(s) \pm \tilde{A}(s))ds + \sum_{k=0}^{m} N_2(A_k \pm \tilde{A}_k).$$

We have

$$N_2(K(i\omega)) \le \sqrt{n}|\omega| + \mathrm{Var}_2(R_0), N_2(K(i\omega) - \tilde{K}(i\omega)) \le \mathrm{Var}_2(R_0 - \tilde{R})$$

and

$$N_2(K(i\omega) + \tilde{K}(i\omega)) \le 2\sqrt{n}|\omega| + \mathrm{Var}_2(R_0 + R) \;\; (\omega \in \mathbb{R}).$$

We thus arrive at the following corollary.

Corollary 5.5.6. *Let K and $\tilde{K}$ be defined by* (5.2). *Then the inequality*

$$\begin{aligned} &|\det\, K(i\omega) - \det\, \tilde{K}(i\omega)| \\ &\le \frac{1}{n^{n/2}} \mathrm{Var}_2(R_0 - \tilde{R}) \left[1 + \sqrt{n}|\omega| + \frac{1}{2}\left(\mathrm{Var}_2(R_0 + \tilde{R}) + \mathrm{Var}_2(R_0 - \tilde{R})\right)\right]^n \end{aligned}$$

is true.

In particular,

$$\begin{aligned} &|\det\, \tilde{K}(i\omega)| \ge |\det\, K(i\omega)| \\ &- \frac{1}{n^{n/2}} \mathrm{Var}_2(R_0 - \tilde{R}) \left[1 + \sqrt{n}|\omega| + \frac{1}{2}\left(\mathrm{Var}_2(R_0 + \tilde{R}) + \mathrm{Var}_2(R_0 - \tilde{R})\right)\right]^n \\ &\qquad (\omega \in \mathbb{R}). \end{aligned}$$

Hence, making use of Lemma 5.5.1 and taking into account that $\mathrm{var}(R_0) \le \mathrm{Var}_2(R_0)$, we easily get our next result.

Corollary 5.5.7. *Let K and $\tilde{K}$ be defined by* (5.2). *Let all the zeros of* $\det\, K(z)$ *be in C_- and*

$$\begin{aligned} J_0 := &\inf_{|\omega| \le 2\mathrm{Var}_2(R_0)} |\det K(i\omega)| \\ &- \frac{1}{n^{n/2}} \mathrm{Var}_2(R_0 - \tilde{R}) \left[1 + 2\sqrt{n}\mathrm{Var}_2(\tilde{R}) + \frac{1}{2}\left(\mathrm{Var}_2(R_0 + \tilde{R}) + \mathrm{Var}_2(R_0 - \tilde{R})\right)\right]^n > 0. \end{aligned}$$

Then all the zeros of $\det\, \tilde{K}(z)$ *are also in C_- and*

$$\inf_{\omega \in \mathbb{R}} |\det\, \tilde{K}(i\omega)| \ge J_0.$$

5.6 Approximations by polynomial pencils

Let (1.2) hold, and K, and K_1 be defined as in Section 5.1. Let us consider the approximation of function K_1 by the polynomial pencil

$$h_m(\lambda) = \sum_{k=0}^{m} \frac{C_k \lambda^k}{k!}.$$

Put

$$q_m(K) := \left[\sum_{k=m+1}^{\infty} \|C_k\|^2 \right]^{1/2}$$

and

$$w(h_m) = 2\, N_2^{1/2} \left(\sum_{k=1}^{m} C_k C_k^* \right) + 2\, [n(\zeta(2)-1)]^{1/2}.$$

Define $\xi(h_m, s)$ by (4.6) with h_m instead of K. The following result at once follows from [46, Corollary 12.5.3].

Corollary 5.6.1. *Let conditions* (1.2) *and* (1.3) *hold, and* $r_m(K)$ *be the unique positive root of the equation*

$$q_m(K)\xi(h_m, y) = 1.$$

Then either, for any characteristic value $z(K)$ *of* K, *there is a characteristic value* $z(h_m)$ *of polynomial pencil* h_m, *such that*

$$\left| \frac{1}{z(K)} - \frac{1}{z(h_m)} \right| \le r_m(K), \quad \text{or} \quad |z(K)| \ge \frac{1}{r_m(K)}.$$

5.7 Convex functions of characteristic values

We need the following classical result.

Lemma 5.7.1. *Let* $\phi(x)$ $(0 \le x \le \infty)$ *be a convex continuous function, such that* $\phi(0) = 0$, *and*

$$a_j, b_j \ \ (j = 1, 2, \ldots, l \le \infty)$$

be two non-increasing sequences of real numbers, such that

$$\sum_{k=1}^{j} a_k \le \sum_{k=1}^{j} b_k \ \ (j = 1, 2, \ldots, l).$$

Then

$$\sum_{k=1}^{j} \phi(a_k) \le \sum_{k=1}^{j} \phi(b_k) \ \ (j = 1, 2, \ldots, l).$$

For the proof see for instance [65, Lemma II.3.4], or [64, p. 53]. Put

$$\chi_k = \omega_k(K) + \frac{n}{k+n} \quad (k = 1, 2, \ldots).$$

The following result is due to the previous lemma and Theorem 5.1.2.

Corollary 5.7.2. *Let $\phi(t)$ $(0 \leq t < \infty)$ be a continuous convex scalar-valued function, such that $\phi(0) = 0$. Let conditions* (1.2) *and* (1.3) *hold. Then the inequalities*

$$\sum_{k=1}^{j} \phi\left(\frac{1}{|z_k(K)|}\right) < \sum_{k=1}^{j} \phi(\chi_k) \quad (j = 1, 2, \ldots)$$

are valid. In particular, for any $p > 1$,

$$\sum_{k=1}^{j} \frac{1}{|z_k(K)|^p} < \sum_{k=1}^{j} \chi_k^p$$

and thus

$$\left[\sum_{k=1}^{j} \frac{1}{|z_k(K)|^p}\right]^{1/p} < \left[\sum_{k=1}^{j} \omega_k^p(K)\right]^{1/p} + n\left[\sum_{k=1}^{j} \frac{1}{(k+n)^p}\right]^{1/p} \quad (j = 1, 2, \ldots). \tag{7.1}$$

For any $p > 1$ we have

$$\zeta_n(p) := \sum_{k=1}^{\infty} \frac{1}{(k+n)^p} < \infty.$$

Relation (7.1) with the notation

$$N_p(\Psi_K) = \left[\sum_{k=1}^{n} \lambda_k^p(\Psi_K)\right]^{1/p}$$

yields our next result.

Corollary 5.7.3. *Let the conditions* (1.2) *and* (1.3) *hold. Then*

$$\left(\sum_{k=1}^{\infty} \frac{1}{|z_k(K)|^p}\right)^{1/p} < N_p(\Psi_K) + n\zeta_n^{1/p}(p) \quad (p > 1).$$

The next result is also well known, cf. [65, Chapter II], [64, p. 53].

Lemma 5.7.4. *Let a scalar-valued function $\Phi(t_1, t_2, \ldots, t_j)$ with an integer j be defined on the domain*

$$0 < t_j \leq t_{j-1} \cdots \leq t_2 \leq t_1 < \infty$$

and have continuous partial derivatives, satisfying the condition

$$\frac{\partial \Phi}{\partial t_1} > \frac{\partial \Phi}{\partial t_2} > \cdots > \frac{\partial \Phi}{\partial t_j} > 0 \quad \textit{for} \quad t_1 > t_2 > \cdots > t_j, \tag{7.2}$$

and a_k, b_k $(k = 1, 2, \ldots, j)$ *be two non-increasing sequences of real numbers satisfying the condition*

$$\sum_{k=1}^{m} a_k \leq \sum_{k=1}^{m} b_k \ (m = 1, 2, \ldots, j).$$

Then $\Phi(a_1, \ldots, a_j) \leq \Phi(b_1, \ldots, b_j)$.

This lemma and Theorem 5.1.2 immediately imply our next result.

Corollary 5.7.5. *Under the hypothesis of Theorem* 5.1.2, *let condition* (7.2) *hold. Then*

$$\Phi\left(\frac{1}{|z_1(K)|}, \frac{1}{|z_2(K)|}, \ldots, \frac{1}{|z_j(K)|}\right) < \Phi(\chi_1, \chi_2, \ldots, \chi_j).$$

In particular, let $\{d_k\}_{k=1}^{\infty}$ be a decreasing sequence of non-negative numbers. Take

$$\Phi(t_1, t_2, \ldots, t_j) = \sum_{k=1}^{j} d_k t_k.$$

Then the previous corollary yields the inequalities

$$\sum_{k=1}^{j} \frac{d_k}{|z_k(K)|} < \sum_{k=1}^{j} \chi_k d_k = \sum_{k=1}^{j} d_k \left[\omega_k(K) + \frac{n}{k+n}\right] \quad (j = 1, 2, \ldots).$$

5.8 Comments

The material of this chapter is adapted from Chapter 12 of the book [46]. The relevant results in more general situations can be found in [32, 33] and [35].

Chapter 6

Equations Close to Autonomous and Ordinary Differential Ones

In this chapter we establish explicit stability conditions for linear time variant systems with delay "close" to ordinary differential systems and for systems with small delays. We also investigate perturbations of autonomous equations.

6.1 Equations "close" to ordinary differential ones

Let $R(t,\tau) = (r_{jk}(t,\tau))_{j,k=1}^{n}$ be an $n\times n$-matrix-valued function defined on $[0,\infty)\times [0,\eta\,]$ $(0 < \eta\ < \infty)$, which is piece-wise continuous in t for each τ, whose entries have bounded variations in τ.

Consider the equation

$$\dot{y}(t) = A(t)y(t) + \int_0^\eta d_\tau R(t,\tau)y(t-\tau) \ \ (t \ge 0), \tag{1.1}$$

where $A(t)$ is a piece-wise continuous matrix-valued function. In this chapter again $C(\Omega) = C(\Omega,\mathbb{C}^n)$ and $L^p(\Omega) = L^p(\Omega,\mathbb{C}^n)$ $(p\ge 1)$ are the spaces of vector-valued functions.

Introduce in $L^1(-\eta\,,\infty)$ the operator

$$Ew(t) = \int_0^\eta d_\tau R(t,\tau)w(t-\tau).$$

It is assumed that

$$v_{jk} = \sup_{t\ge 0} \operatorname{var}(r_{jk}(t,.)) < \infty. \tag{1.2}$$

By Lemma 1.12.3, there is a constant

$$q_1 \le \sum_{j=1}^{n} \sqrt{\sum_{k=1}^{n} v_{jk}^2}$$

such that

$$\|Ew\|_{L^1(0,\infty)} \le q_1 \|w\|_{L^1(-\eta\,,\infty)} \quad (w \in L^1(-\eta\,,\infty)). \tag{1.3}$$

For instance, if (1.1) takes the form

$$\dot{y}(t) = A(t)y + \int_0^{\eta} B(t,s)y(t-s)ds + \sum_{k=0}^{m} B_k(t)y(t-\tau_k) \;\; (t \ge 0;\; m < \infty), \tag{1.4}$$

where $0 \le \tau_0 < \tau_1, \dots, < \tau_m \le \eta$ are constants, $B_k(t)$ are piece-wise continuous matrices and $B(t,s)$ is a matrix function Lebesgue integrable in s on $[0, \eta\,]$, then (1.3) holds, provided

$$\hat{q}_1 := \sup_{t \ge 0} \left(\int_0^{\eta} \|B(t,s)\|_n ds + \sum_{k=0}^{m} \|B_k(t)\|_n \right) < \infty.$$

Here and below in this chapter $\|A\|_n$ is the spectral norm of an $n \times n$-matrix A. Moreover, we have

$$\int_0^{\infty} \|Ef(t)\|_n dt \le \int_0^{\infty} \left(\int_0^{\eta} \|B(t,s)f(t-s)\|_n ds + \sum_{k=0}^{m} \|B_k(t)f(t-\tau_k)\|_n \right) dt$$
$$\le \sup_{\tau} \int_0^{\eta} \|B(\tau,s)\|_n \int_0^{\infty} \|f(t-s)\|_n ds\; dt + \sum_{k=0}^{m} \|B_k(\tau)\|_n \int_0^{\infty} \|f(t-\tau_k)\|_n dt.$$

Consequently,

$$\int_0^{\infty} \|Ef(t)\|_n dt \le \hat{q}_1(R) \left(\max_{0 \le s \le \eta} \int_0^{\infty} \|f(t-s)\|_n dt + \int_0^{\infty} \sum_{k=0}^{m} \|f(t-\tau_k)\|_n dt \right).$$

But

$$\max_{0 \le s \le \eta} \int_0^{\infty} \|f(t-s)\|_n dt \le \int_{-\eta}^{\infty} \|f(t)\|_n dt.$$

Thus, in the case of equation (1.4), condition (1.3) holds with $q_1 = \hat{q}_1$.

Theorem 6.1.1. *Let condition* (1.3) *hold and the evolution operator* $U(t,s)$ $(t \ge s \ge 0)$ *of the equation*

$$\dot{y} = A(t)y \;(t > 0) \tag{1.5}$$

satisfy the inequality

$$\nu_1 := \sup_{s \ge 0} \int_s^{\infty} \|U(t,s)\|_n dt < \frac{1}{q_1}. \tag{1.6}$$

Then equation (1.1) *is exponentially stable.*

To prove this theorem we need the following result.

Lemma 6.1.2. *Let conditions* (1.3) *and* (1.6) *hold. Then a solution* $x(t)$ *of the equation*

$$\dot{x}(t) = A(t)x(t) + \int_0^\eta d_\tau R(t,\tau)x(t-\tau) + f(t) \tag{1.7}$$

with an $f \in L^1(0,\infty)$ *and the zero initial condition*

$$x(t) = 0 \ \ (t \le 0) \tag{1.8}$$

satisfies the inequality

$$\|x\|_{L^1(0,\infty)} \le \frac{\nu_1 \|f\|_{L^1(0,\infty)}}{1-\nu_1 q_1}.$$

Proof. Equation (1.1) is equivalent to the following one:

$$x(t) = \int_0^t U(t,s)(Ex(s) + f(s))ds.$$

So

$$\|x(t)\|_n \le \int_0^t \|U(t,s)\|_n(\|Ex(s)\|_n + \|f(s)\|_n)ds.$$

Integrating this inequality, we obtain

$$\int_0^{t_0} \|x(t)\|_n dt \le \int_0^{t_0}\int_0^t \|U(t,s)\|_n(\|Ex(s)\|_n + \|f(s)\|_n)ds\, dt \ \ (0 < t_0 < \infty).$$

Take into account that

$$\int_0^{t_0}\int_0^t \|U(t,s)\|_n \|f(s)\|_n ds\, dt = \int_0^{t_0} \|f(s)\|_n \int_s^{t_0} \|U(t,s)\|_n dt\, ds \le \nu_1 \|f\|_{L^1(0,\infty)}.$$

In addition,

$$\int_0^{t_0}\int_0^t \|U(t,s)\|_n \|Ex(s)\|_n ds\, dt = \int_0^{t_0} \|Ex(s)\|_n \int_s^{t_0} \|U(t,s)\|_n dt\, ds$$
$$\le q_1\nu_1 \int_0^{t_0} \|x(s)\|_n ds.$$

Thus,

$$\int_0^{t_0} \|x(s)\|_n ds \le \nu_1 q_1 \int_0^{t_0} \|x(s)\|_n ds + \nu_1 \|f\|_{L^1(0,\infty)}.$$

Hence,

$$\int_0^{t_0} \|x(s)\|_n ds \le \frac{\nu_1 \|f\|_{L^1(0,t_0)}}{1-\nu_1 q_1}.$$

Now letting $t_0 \to \infty$, we arrive at the required result. □

The assertion of Theorem 6.1.1 is due to Theorem 3.5.1 and the previous lemma.

Now consider the operator E in space C. By Lemma 1.12.3, under condition (1.2) there is a constant

$$q_\infty \le \sqrt{n} \| \left(v_{jk}\right)_{j,k=1}^n \|_n,$$

such that

$$\|Ew\|_{C(0,\infty)} \le q_\infty \|w\|_{C(-\eta\,,\infty)} \quad (w \in C(-\eta\,,\infty)). \tag{1.9}$$

For instance, if (1.1) takes the form

$$\dot{y}(t) = A(t)y + \int_0^\eta B(t,s)y(t-s)ds + \sum_{k=1}^m B_k(t)y(t-h_k(t)) \quad (t \ge 0;\ m < \infty), \tag{1.10}$$

where $0 \le h_1(t), h_2(t), \dots, h_m(t) \le \eta$ are continuous functions, $B(t,s)$ and $B_k(t)$ are the same as above in this section. Then

$$\sup_{t\ge 0} \|Ew(t)\|_n \le \sup_{t \ge 0} \left(\int_0^\eta \|B(t,s)w(t-s)\|_n ds + \sum_{k=0}^m \|B_k(t)w(t-h_k(t))\|_n \right).$$

Hence, under the condition $\hat{q}_1 < \infty$, we easily obtain inequality (1.9) with $q_\infty = \hat{q}_1$.

Theorem 6.1.3. *Let condition* (1.9) *hold and the evolution operator* $U(t,s)$ $(t \ge s \ge 0)$ *of equation* (1.5) *satisfy the inequality*

$$\nu_\infty := \sup_{t\ge 0} \int_0^t \|U(t,s)\|_n ds < \frac{1}{q_\infty}. \tag{1.11}$$

Then equation (1.1) *is exponentially stable.*

The assertion of this theorem follows from Theorem 3.4.1 and the following lemma.

Lemma 6.1.4. *Let conditions* (1.9) *and* (1.11) *hold. Then a solution* $x(t)$*of problem* (1.7), (1.8) *with a* $f \in C(0,\infty)$ *satisfies the inequality*

$$\|x\|_{C(0,\infty)} \le \frac{\nu_\infty \|f\|_{C(0,\infty)}}{1 - \nu_\infty q_\infty}.$$

The proof of this lemma is similar to the proof of Lemma 6.1.2.

Assume that

$$((A(t) + A^*(t))h, h)_{C^n} \le -2\alpha(t)(h,h)_{C^n} \quad (h \in \mathbb{C}^n, t \ge 0)$$

with a positive piece-wise continuous function $\alpha(t)$ having the property

$$\hat{\nu}_1 := \sup_{s \ge 0} \int_s^\infty e^{-\int_s^t \alpha(t_1)dt_1} dt < \infty.$$

Then we easily have the inequalities

$$\|U(t,s)\|_n \le e^{-\int_s^t \alpha(t_1)dt_1}$$

and $\nu_1 \le \hat{\nu}_1$. For instance, if $\alpha_0 := \inf_t \alpha(t) > 0$, then

$$\hat{\nu}_1 \le \sup_{s\ge 0} \int_s^\infty e^{-\alpha_0(t-s)}dt = \frac{1}{\alpha_0}.$$

Let $A(t) \equiv A_0$-a constant matrix. Then

$$U(t,s) = e^{A_0(t-s)} \quad (t \ge s \ge 0).$$

In this case

$$\nu_1(A) = \nu_\infty(A) = \|e^{A_0 t}\|_{L^1(0,\infty)}.$$

Applying Corollary 2.5.3, we have

$$\|e^{At}\|_n \le e^{\alpha(A)t} \sum_{k=0}^{n-1} \frac{g^k(A)t^k}{(k!)^{3/2}} \quad (t \ge 0).$$

Recall that

$$g(A) = (N^2(A) - \sum_{k=1}^n |\lambda_k(A)|^2)^{1/2} \le \sqrt{2}N(A_I),$$

and $\alpha(A) = \max_k \operatorname{Re} \lambda_k(A)$; $\lambda_k(A)$ $(k = 1, \dots, n)$ are the eigenvalues of A, $N_2(A)$ is the Hilbert–Schmidt norm of A and $A_I = (A - A^*)/2i$. Thus

$$\|e^{A_0 t}\|_{L^1(0,\infty)} \le \nu_{A_0}, \quad \text{where} \quad \nu_{A_0} := \sum_{k=0}^{n-1} \frac{g^k(A_0)}{\sqrt{k!}|\alpha(A_0)|^{k+1}}.$$

6.2 Equations with small delays

Again consider the equation

$$\dot{y}(t) = \int_0^\eta d_\tau R(t,\tau)y(t-\tau) \quad (t \ge 0), \tag{2.1}$$

where $R(t,\tau) = (r_{jk}(t,\tau))$ is the same as in the previous section. In particular, condition (1.2) holds.

For instance, (2.1) can take the form

$$\dot{y}(t) = \int_0^\eta B(t,s)y(t-s)ds + \sum_{k=0}^m B_k(t)y(t-\tau_k) \quad (t \ge 0;\ m < \infty) \tag{2.2}$$

where $B(t,s), \tau_k$ and $B(t,\tau)$ are the same as in the previous section. In $C(-\eta,\infty)$ introduce the operator

$$\hat{E}_d f(t) := \int_0^\eta d_\tau R(t,\tau)[f(t-\tau) - f(t)].$$

Assume that there is a constant $\tilde{V}_d(R)$, such that

$$\|\hat{E}_d f\|_{C(0,T)} \le \tilde{V}_d(R)\|\dot{f}\|_{C(-\eta,T)} \quad (T > 0) \tag{2.3}$$

for any $f \in C(-\eta,T)$ with $\dot{f} \in C(-\eta,T)$.

It is not not hard to show that condition (1.2) implies (2.3).

In the case of equation (2.2) we have

$$\hat{E}_d f(t) = \int_0^\eta B(t,\tau)[f(t-\tau) - f(t)]d\tau + \sum_{k=0}^m B_k(t)[f(t-\tau_k) - f(t)].$$

Since

$$\|f(t-\tau) - f(t)\|_{C(0,T)} \le \tau\|\dot{f}\|_{C(-\eta,T)},$$

we easily obtain

$$\|\hat{E}_d f(t)\|_{C(0,T)} \le \sup_t \left(\int_0^\eta \tau\|B(t,\tau)\|_n + \sum_{k=0}^m \|B_k(t)\|_n \tau_k\right) \|\dot{f}\|_{C(-\eta,T)}.$$

That is, for equation (2.2), condition (2.3) holds with

$$\tilde{V}_d(R) \le \sup_t \left(\int_0^\eta \tau\|B(t,\tau)\|_n + \sum_{k=0}^m \tau_k\|B_k(t)\|_n\right). \tag{2.4}$$

Put

$$A(t) = R(t,\eta) - R(t,0) \tag{2.5}$$

and assume that equation (1.5) is asymptotically stable. Recall that $U(t,s)$ is the evolution operator of the ordinary differential equation (1.5).

Theorem 6.2.1. *Under conditions* (1.2) *and* (2.3), *let* $A(t)$ *be defined by* (2.5) *and*

$$\psi_R := \tilde{V}_d(R)\left(\sup_{t\ge 0}\int_0^t \|A(t)U(t,s)\|_n ds + 1\right) < 1. \tag{2.6}$$

Then equation (2.1) *is exponentially stable.*

Proof. We need the non-homogeneous problem

$$\dot{x}(t) = Ex(t) + f \quad (t \ge 0;\ f \in C(0,\infty)), \tag{2.7}$$

with the zero initial condition

$$x(t) = 0, t \leq 0.$$

Observe that

$$\begin{aligned}\int_0^{\eta} d_\tau R(t,\tau)x(t-\tau) &= \int_0^{\eta} d_\tau R(t,\tau)x(t) + \int_0^{\eta} d_\tau R(t,\tau)(x(t-\tau) - x(t)) \\ &= (R(t,\eta) - R(t,0))x(t) + \int_0^{\eta} d_\tau R(t,\tau)(x(t-\tau) - x(t)) \\ &= A(t)x(t) + \hat{E}_d x(t).\end{aligned}$$

So we can rewrite equation (2.7) as

$$\dot{x}(t) = A(t)x(t) + \hat{E}_d x(t) + f(t) \quad (t \geq 0). \tag{2.8}$$

Consequently,

$$x(t) = \int_0^t U(t,s)\hat{E}_d x(s)ds + f_1(t), \tag{2.9}$$

where

$$f_1(t) = \int_0^t U(t,s)f(s)ds.$$

Differentiating (2.9), we get

$$\dot{x}(t) = \int_0^t A(t)U(t,s)\hat{E}_d x(s)ds + \hat{E}_d x(t) + A(t)f_1(t) + f(t).$$

For brevity put $|x|_T = \|x\|_{C(0,T)}$ for a finite T. By condition (2.3) we have

$$|\dot{x}|_T \leq c_0 + \tilde{V}_d(R)|\dot{x}|_T \left(\sup_{t\geq 0} \int_0^t \|A(t)U(t,s)\|_n ds + 1 \right),$$

where

$$c_0 := \|A(t)f_1\|_{C(0,\infty)} + \|f\|_{C(0,\infty)}.$$

So

$$|\dot{x}|_T \leq c_0 + \psi_R |\dot{x}|_T.$$

Now condition (2.6) implies

$$|\dot{x}|_T \leq \frac{c_0}{1-\psi_R}.$$

Letting $T \to \infty$, we can assert that $\dot{x} \in C(0,\infty)$. Hence due to (2.3) and (2.9) it follows that $x(t)$ is bounded, since

$$\|x\|_{C(0,\infty)} \leq \|f_1\|_{C(0,\infty)} + \tilde{V}_d(R)\|\dot{x}\|_{C(0,\infty)} \sup_{t\geq 0} \int_0^t \|U(t,s)\|_n ds.$$

Now the required result is due to Theorem 3.4.1. $\square$

6.3 Nonautomomous systems "close" to autonomous ones

In the present section we consider the vector equation

$$\dot{y}(t) = \int_0^{\eta} dR_0(\tau)y(t-\tau) + \int_0^{\eta} d_\tau R(t,\tau)y(t-\tau) = 0 \ \ (t \geq 0), \tag{3.1}$$

where $R(t,\tau)$ is the same as in Section 6.1, and $R_0(\tau) = (r_{jk}^{(0)}(\tau))_{j,k=1}^n$ is an $n \times n$-matrix-valued function defined on $[0,\eta\,]$, whose entries have bounded variations. Recall that $\mathrm{var}(R_0)$ is the spectral norm of the matrix

$$\left(\mathrm{var}(r_{jk}^{(0)})\right)_{j,k=1}^n$$

(see Section 1.12). Again use the operator $Ef(t) = \int_0^\eta d_\tau R(t,\tau)f(t-\tau)$, considering it in space $L^2(-\eta\,,\infty)$.

Under condition (1.2), due to Lemma 1.12.3, there is a constant

$$q_2 \leq \|\,(v_{jk})_{j,k=1}^n\,\|_n,$$

such that

$$\|Ef\|_{L^2(0,\infty)} \leq q_2\|f\|_{L^2(-\eta\,,\infty)} \ \ (f \in L^2(-\eta\,,\infty)). \tag{3.2}$$

Again

$$K(z) = zI - \int_0^{\eta} \exp(-zs)dR_0(s) \ \ (z \in \mathbb{C})$$

is the characteristic matrix of the autonomous equation

$$\dot{y}(t) = \int_0^{\eta} dR_0(\tau)y(t-\tau) \ \ (t \geq 0). \tag{3.3}$$

As above it is assumed that all the zeros of $\det\ K(z)$ are in the open left half-plane. Recall also that

$$\theta(K) := \sup_{-2\ \mathrm{var}(R_0)\leq\omega\leq 2\ \mathrm{var}(R_0)} \|K^{-1}(i\omega)\|_n,$$

and in Chapter 4, estimates for $\theta(K)$ are given.

The equation

$$\begin{aligned}\dot{y}(t) = \int_0^{\eta} A(s)y(t-s)ds + \sum_{k=0}^{m} A_k y(t-\tau_k) + \int_0^{\eta} B(t,s)y(t-s)ds \\ + \sum_{k=0}^{\tilde{m}} B_k(t)y(t-h_k) \ \ (t \geq 0;\ m, \tilde{m} < \infty)\end{aligned} \tag{3.4}$$

is an example of equation (3.1). Here

$$0 = h_0 < h_1 < \cdots < h_{\tilde{m}} \leq \eta \quad \text{and} \quad 0 = \tau_0 < \tau_1 < \cdots < \tau_m \leq \eta$$

are constants, A_k are constant matrices and $A(s)$ is Lebesgue integrable on $[0, \eta\,]$, $B_k(t)$ are piece-wise continuous matrices and $B(t,s)$ is Lebesgue integrable in s on $[0, \eta\,]$. In this case,

$$K(z) = zI - \int_0^\eta e^{-sz} A(s)ds - \sum_{k=0}^m e^{-h_k z} A_k \tag{3.5}$$

and

$$\operatorname{var}(R_0) \leq \int_0^\eta \|A(s)\|_n ds + \sum_{k=0}^m \|A_k\|_n. \tag{3.6}$$

Theorem 6.3.1. *Let the conditions* (3.2) *and*

$$q_2 \theta(K) < 1 \tag{3.7}$$

be fulfilled. Then equation (3.1) *is exponentially stable.*

Proof. Consider the non-homogeneous equation

$$\dot{x} = E_0 x + E x + f, \tag{3.8}$$

with the zero initial condition and $f \in L^2(0, \infty)$. Here

$$E_0 x(t) = \int_0^\eta dR_0(\tau) x(t - \tau).$$

By Lemma 4.4.1 and (3.2),

$$\|x\|_{L^2(0,\infty)} \leq \theta(K) \|Ex + f\|_{L^2(0,\infty)}.$$

Thus, by (3.2),

$$\|x\|_{L^2(0,\infty)} \leq \theta(K)(q_2 \|x\|_{L^2(0,\infty)} + \|f\|_{L^2(0,\infty)}).$$

Hence, taking into account (3.7), we get

$$\|x\|_{L^2(0,\infty)} \leq \frac{\theta(K)\|f\|_{L^2(0,\infty)}}{1 - \theta(K) q_2}. \tag{3.9}$$

Now Theorem 3.5.1 implies the required result. □

Example 6.3.2. Consider the equation

$$\dot{y}(t) + A_1 \int_0^\eta y(t-s) dr_1(s) + A_2 \int_0^\eta y(t-s) dr_2(s) = (Ey)(t) \tag{3.10}$$

with commuting positive definite Hermitian matrices A_1, A_2, and scalar nondecreasing functions $r_1(s), r_2(s)$.

Besides,

$$K(z) = zI + A_1 \int_0^\eta e^{-zs} dr_1(s) + A_2 \int_0^\eta e^{-zs} dr_2(s).$$

In addition, the eigenvalues $\lambda_j(A_1), \lambda_j(A_2)$ $(j = 1, \ldots, n)$ of A_1 and A_2, respectively are positive. Moreover, since the matrices commute, one can enumerate the eigenvalues in such a way, that the eigenvalues of $K(z)$ are defined by

$$\lambda_j(K(z)) = z + \lambda_j(A_1) \int_0^\eta e^{-zs} dr_1(s) + \lambda_j(A_2) \int_0^\eta e^{-zs} dr_2(s) \quad (j = 1, \ldots, n).$$

In addition,

$$\operatorname{var}(R_0) \le \|A_1\|_n \operatorname{var}(r_1) + \|A_2\|_n \operatorname{var}(r_2).$$

Let us use Corollary 4.3.2. In the considered case $B(z)$ is normal. So $g(B(z)) \equiv 0$ and thus the mentioned corollary implies

$$\theta(K) \le \frac{1}{\hat{d}(K)}$$

with

$$\hat{d}(K) = \min_{j=1,\ldots,n} \inf_{|\omega| \le 2 \operatorname{var}(R_0)} \left| i\omega + \lambda_j(A_1) \int_0^\eta e^{-i\omega s} dr_1(s) + \lambda_j(A_2) \int_0^\eta e^{-i\omega s} dr_2(s) \right|.$$

Letting

$$\tilde{r}_j(s) = \lambda_j(A_1) r_1(s) + \lambda_j(A_2) r_2(s),$$

we obtain

$$\operatorname{var}(\hat{r}_j) = \lambda_j(A_1) \operatorname{var}(r_1) + \lambda_j(A_2) \operatorname{var}(r_2)$$

and

$$\lambda_j(K(z)) = z + \int_0^\eta e^{-zs} d\hat{r}_j(s) \quad (j = 1, \ldots, n).$$

Assume that

$$e\eta \ \operatorname{var}(\hat{r}_j) < 1, \quad j = 1, \ldots, n. \tag{3.11}$$

Then by Lemma 4.6.5, we obtain

$$\hat{d}(K) = \min_j \operatorname{var}(\hat{r}_j).$$

Applying Theorem 6.3.1 we can assert that equation (3.10) is exponentially stable, provided the conditions (3.2), (3.11) and

$$q_2 < \min_j \operatorname{var}(\hat{r}_j) = \min_j (\lambda_j(A_1) \operatorname{var}(r_1) + \lambda_j(A_2) \operatorname{var}(r_2))$$

hold.

6.4 Equations with constant coefficients and variable delays

Consider in $\mathbb{C}^n$ the equation

$$\dot{x}(t) = \sum_{k=1}^{m} A_k x(t - \tau_k(t)) + f(t) \quad (f \in L^2(0,\infty);\ t \geq 0) \tag{4.1}$$

with condition (1.8). Here A_k are constant matrices, $\tau_k(t)$ are non-negative continuous scalar functions defined on $[0,\infty)$ and satisfying the conditions

$$h_k \leq \tau_k(t) \leq \eta_k \quad (0 \leq h_k, \eta_k \equiv \text{const} \leq \eta;\ k = 1,\dots,m < \infty;\ t \geq 0). \tag{4.2}$$

Introduce the matrix function

$$K(z) = zI - \sum_{k=1}^{m} A_k e^{-zh_k} \quad (z \in \mathbb{C}).$$

As above, it is assumed that *all the roots of* $\det\ K(z)$ *are in* C_-. Set

$$v_0 := \sum_{k=1}^{m} \|A_k\|_n,\ \theta(K) := \max_{-2v_0 \leq s \leq 2v_0} \|K^{-1}(is)\|_n$$

and

$$\gamma(K) := \sum_{k=1}^{m} (\eta_k - h_k)\|A_k\|_n.$$

Theorem 6.4.1. *Let the conditions* (4.2) *and*

$$v_0\theta(K) + \gamma(K) < 1 \tag{4.3}$$

hold. Then a solution of equation (4.1) *with the zero initial condition* (1.8) *satisfies the inequality*

$$\|x\|_{L^2(0,\infty)} \leq \frac{\theta(K)\|f\|_{L^2(0,\infty)}}{1 - v_0\theta(K) - \gamma(K)}. \tag{4.4}$$

This theorem is proved in the next section.

Theorems 6.4.1 and 3.5.1 imply

Corollary 6.4.2. *Let conditions* (4.2) *and* (4.3) *hold. Then the equation*

$$\dot{y}(t) = \sum_{k=1}^{m} A_k y(t - \tau_k(t)) \tag{4.5}$$

is exponentially stable.

For instance, consider the following equation with one delay:

$$\dot{y}(t) = A_0 y(t - \tau(t)) \quad (t > 0), \tag{4.6}$$

where A_0 is a constant $n \times n$-matrix and the condition

$$h \le \tau(t) \le \eta \quad (h \equiv \text{const} \ge 0;\ t \ge 0) \tag{4.7}$$

holds. In the considered case $K(z) = zI - e^{-zh}A_0$. As it is shown in Section 4.3, for any regular z,

$$\|K^{-1}(z)\|_n \le \Gamma(K(z)) \quad (z \notin \Sigma(K))$$

where

$$\Gamma(K(z)) = \sum_{k=0}^{n-1} \frac{g^k(B(z))}{\sqrt{k!}\,d^{k+1}(K(z))},$$

with $B(z) = e^{-zh}A_0$ and $d(K(z))$ is the smallest modulus of eigenvalues of $K(z)$:

$$d(K(z)) = \min_{k=1,\dots,n} |\lambda_k(K(z))|.$$

Here

$$\lambda_j(K(z)) = z - e^{-zh}\lambda_j(A_0)$$

are the eigenvalues of matrix $K(z)$ counting with their multiplicities.

For a real ω we have $g(B(i\omega)) = g(A_0)$. In addition,

$$v_0 = \|A_0\|_n \quad \text{and} \quad \gamma(K) := (\eta - h)\|A_0\|_n.$$

Thus

$$\theta(K) \le \Gamma_0(K)) := \max_{|\omega| \le 2v_0} \Gamma(K(i\omega)) \le \theta_{A_0},$$

where

$$\theta_{A_0} := \sum_{k=0}^{n-1} \frac{g^k(A_0)}{\sqrt{k!}\,d^{k+1}(K)},$$

and

$$d(K) := \inf_{j=1,\dots,n;\ |y| \le 2|\lambda_j(A_0)|} |yi + \lambda_j(A_0)e^{-iyh}|.$$

In particular, if A_0 is a normal matrix, then $g(A_0) = 0$ and $\theta_A = 1/d(K)$.

Now Theorem 6.4.1 yields the following result.

Corollary 6.4.3. *Let the conditions* (4.7) *and*

$$v_0\theta_{A_0} + (\eta - h)\|A_0\|_n < 1 \tag{4.8}$$

hold. Then equation (4.6) *is exponentially stable.*

6.5 Proof of Theorem 6.4.1

To prove Theorem 6.4.1 we need the following result.

Lemma 6.5.1. *Let $\tau(t)$ be a non-negative continuous scalar function defined on $[0,\infty)$ and satisfying condition (4.7). In addition, let a function $w \in L^2(-\eta,\infty)$ have the properties $\dot{w} \in L^2(0,\infty)$ and $w(t) = 0$ for $t < 0$. Then*

$$\|w(t-\tau(t)) - w(t-h)\|_{L^2(0,\infty)} \leq (\eta - h)\|\dot{w}\|_{L^2(0,\infty)}.$$

Proof. Put

$$u(t) = w(t-h) - w(t-\tau(t)) = \int_{t-\tau(t)}^{t-h} \dot{w}(s)ds.$$

By the Schwarz inequality and (4.7) we obtain,

$$\begin{aligned}
\|u\|^2_{L^2(0,\infty)} &= \int_0^\infty \| \int_{t-\tau(t)}^{t-h} \dot{w}(s)ds\|_n^2 \, dt \\
&\leq \int_0^\infty (\eta - h) \int_{t-\eta}^{t-h} \|\dot{w}(s)\|_n^2 ds \, dt \\
&= \int_0^\infty (\eta - h) \int_0^{\eta-h} \|\dot{w}(t-h-s_1)\|_n^2 ds_1 \, dt \\
&\leq (\eta - h) \int_0^{\eta-h} \int_0^\infty \|\dot{w}(t-s_1-h)\|_n^2 dt \, ds_1 \\
&= (\eta - h) \int_0^{\eta-h} \int_{\eta-s_1-h}^\infty \|\dot{w}(t_1)\|_n^2 dt_1 \, ds_1 \\
&\leq (\eta - h)^2 \int_0^\infty \|\dot{w}(t_1)\|_n^2 dt_1.
\end{aligned}$$

We thus get the required result. □

Now let $w \in L^2(0,T)$ for a sufficiently large finite finite $T > 0$. Extend it to the whole positive half-line by

$$w(t) = w(T)(t - T + 1) \ \ (T < t \leq T+1) \quad \text{and} \quad w(t) = 0 \ \ (t > T+1).$$

Then by the previous lemma we get our next result.

Corollary 6.5.2. *Under condition (4.7) let a function $w \in L^2(-\eta, T)$ $(0 < T < \infty)$ have the properties $\dot{w} \in L^2(0,T)$ and $w(t) = 0$ for $t < 0$. Then*

$$\|w(t-\tau(t)) - w(t-h)\|_{L^2(0,T)} \leq (\eta - h)\|\dot{w}\|_{L^2(0,T)}.$$

Proof of Theorem 6.4.1. In this proof, for brevity we put $\|.\|_{L^2(0,T)} = |.|_T$ $(T < \infty)$. Recall that it is supposed that the characteristic values of K are in the open left half-plane.

From (4.1) it follows that

$$\dot{x}(t) = \sum_{k=1}^{m} A_k x(t - h_k) + [F_0 x](t) \ (t \geq 0), \tag{5.1}$$

where

$$[F_0 x](t) = f(t) + \sum_{k=1}^{m} A_k [x(t - \tau_k(t)) - x(t - h_k)].$$

It is simple to check that by (1.5)

$$|\sum_{k=1}^{m} A_k x(t - h_k)|_T \leq v_0 |x|_T.$$

Moreover, thanks to Corollary 6.5.2,

$$|\sum_{k=1}^{m} A_k [x(t - \tau_k(t)) - x(t - h_k)]|_T \leq \gamma(K) |\dot{x}|_T.$$

So

$$|F_0 x|_T \leq \gamma(K) |\dot{x}|_T + |f|_T \tag{5.2}$$

and

$$|\dot{x}|_T \leq v_0 |x|_T + \gamma(K) |\dot{x}|_T + |f|_T.$$

By (4.3) we have $\gamma(K) < 1$. Hence,

$$|\dot{x}|_T \leq \frac{v_0 |x|_T + |f|_T}{1 - \gamma(K)}. \tag{5.3}$$

Furthermore, by the Variation of Constants formula,

$$x(t) = \int_0^t G(t - s) F_0(s) ds$$

where G is the fundamental solution of the equation

$$\dot{x}(t) = \sum_{k=1}^{m} A_k x(t - h_k).$$

So $|x|_T \leq |\hat{G}|_T |F_0|_T$, where

$$\hat{G} f(t) = \int_0^t G(t - s) f(s) ds.$$

But $|\hat{G}|_T \leq \|\hat{G}\|_{L^2(0,\infty)}$ and according to Lemma 4.4.1,

$$\|\hat{G}\|_{L^2(0,\infty)} = \theta(K).$$

Therefore, from (5.1) it follows that $|x|_T \leq \theta(K)|F_0|_T$. Hence, due to (5.2)

$$|x|_T \leq \theta(K)(\gamma(K)|\dot{x}|_T + |f|_T).$$

Now (5.3) implies

$$|x|_T \leq \theta(K)(\gamma(K)|\dot{x}|_T + |f|_T) \leq \frac{\theta(K)(v_0|x|_T + |f|_T)}{1-\gamma(K)}.$$

By condition (4.3) we have

$$\frac{v_0\theta(K)}{1-\gamma(K)} < 1.$$

Consequently,

$$|x|_T \leq \frac{\theta(K)\|f\|_{L^2(0,\infty)}}{1 - v_0\theta(K) - \gamma(K)}.$$

Letting in this inequality $T \to \infty$, we get the required inequality, as claimed. □

6.6 The fundamental solution of equation (4.1)

Again consider equation (4.1). Denote its fundamental solution by $W(t,s)$ and put

$$\psi(K) = \frac{\theta(K)}{1 - v_0\theta(K) - \gamma(K)},$$

$$A = \sum_{k=1}^{m} A_k \quad \text{and} \quad \chi_A := \max_{0\leq t\leq \eta} \|e^{-At} - I\|_n. \tag{6.1}$$

Theorem 6.6.1. *Under the above notation, let A be a Hurwitz matrix and conditions (4.2) and (4.3) hold. Then for all $s \geq 0$, the following relations are valid:*

$$\|W(.,s)\|_{L^2(s,\infty)} \leq (1 + \chi_A v_0 \psi(K))\|e^{At}\|_{L^2(s,\infty)}, \tag{6.2}$$

$$\|W_t'(.,s)\|_{L^2(s,\infty)} \leq \frac{v_0(1 + v_0\chi_A\psi(K))}{1-\gamma(K)}\|e^{At}\|_{L^2(s,\infty)} \tag{6.3}$$

and

$$\|W(.,s)\|_{C(s,\infty)} \leq \frac{\sqrt{2v_0}\|e^{At}\|_{L^2(s,\infty)}}{\sqrt{1-\gamma(K)}}(1 + \psi(K)\chi_A v_0). \tag{6.4}$$

This theorem is proved in the next section.

Let $f(z)$ be a function holomorphic on a neighborhood of the closed convex hull co(A) of the spectrum of an $n \times n$-matrix and A. Then by Corollary 2.5.3 we have

$$\|f(A)\| \leq \sum_{k=0}^{n-1} \sup_{\lambda\in \text{co}(A)} |f^{(k)}(\lambda)| \frac{g^k(A)}{(k!)^{3/2}}.$$

Hence, in particular,

$$\|e^{At}\|_n \le e^{\alpha(A)t} \sum_{k=0}^{n-1} \frac{g^k(A)t^k}{(k!)^{3/2}} \quad (t \ge 0),$$

and

$$\|e^{-At} - I\|_n \le e^{-\beta(A)t} \sum_{k=1}^{n-1} \frac{g^k(A)t^k}{(k!)^{3/2}} \quad (t \ge 0), \tag{6.5}$$

where

$$\alpha(A) := \max_k \operatorname{Re} \lambda_k(A) \quad \text{and} \quad \beta(A) := \min_k \operatorname{Re} \lambda_k(A).$$

So

$$\|e^{At}\|_{L^2(0,\infty)} \le \left[\int_0^\infty e^{2\alpha(A)t} \left(\sum_{k=0}^{n-1} \frac{g^k(A)t^k}{(k!)^{3/2}} \right)^2 dt \right]^{1/2}. \tag{6.6}$$

This integral is easily calculated. Due to (6.5)

$$\chi_A \le e^{-\beta(A)\eta} \sum_{k=1}^{n-1} \frac{g^k(A)\eta^k}{(k!)^{3/2}}, \tag{6.7}$$

since A is Hurwitzian.

In the rest of this section we illustrate Theorem 6.6.1 in the case of equation (4.6) with one delay. In this case

$$\gamma(K) = (\eta - h)\|A_0\|_n.$$

To estimate χ_{A_0} and $\|e^{A_0 t}\|_{L^2(s,\infty)}$ we can directly apply (6.6) and (6.7). As it is shown in Section 6.4, we have the inequality $\psi(K) \le \psi_{A_0}$, where

$$\psi_{A_0} = \frac{\theta_{A_0}}{1 - \|A_0\|_n \theta_{A_0} - (\eta - h)\|A_0\|_n}.$$

Now Theorem 6.6.1 implies

Corollary 6.6.2. *Let A_0 be a Hurwitz matrix and conditions (4.7), and (4.8) hold. Then for all $s \ge 0$, the fundamental solution $W(.,.)$ of equation (4.6) satisfies the inequalities*

$$\|W(.,s)\|_{L^2(s,\infty)} \le (1 + \psi_{A_0}\|A_0\|_n \chi_{A_0})\|e^{A_0 t}\|_{L^2(s,\infty)},$$

$$\|W_t'(.,s)\|_{L^2(s,\infty)} \le \frac{\|A_0\|_n(1 + \psi_{A_0}\|A\|_n \chi_{A_0})}{1 - (\eta - h)\|A_0\|_n}\|e^{A_0 t}\|_{L^2(s,\infty)}$$

and

$$\|W(.,s)\|^2_{C(s,\infty)} \le \frac{2\|A_0\|_n}{1 - (\eta - h)\|A_0\|_n}(1 + \psi_{A_0}\|A_0\|_n \chi_{A_0})^2 \|e^{A_0 t}\|^2_{L^2(s,\infty)}.$$

Let us use Lemma 4.6.2 which asserts that, for constants $h, a_0 \in (0, \infty)$, we have

$$\inf_{\omega \in \mathbb{R}} |i\omega + a_0 e^{-ih\omega}| \geq \cos(2ha_0) > 0.$$

provided $a_0 h < \phi/4$. From the latter result it follows that, if all the eigenvalues of A_0 are real, negative, and

$$h|\lambda_j(A_0)| < \pi/4 \ \ (j = 1, \dots, n), \tag{6.8}$$

then

$$d(K) \geq \tilde{d}_{A_0} := \min_{j=1,\dots,n} |\lambda_j(A_0)| \ \cos(2h\lambda_j(A)).$$

So under condition (6.8) one has $d(K) \geq \tilde{d}_A$ and therefore,

$$\theta_{A_0} < \sum_{k=0}^{n-1} \frac{g^k(A_0)}{\sqrt{k!}\tilde{d}_A^{k+1}}.$$

6.7 Proof of Theorem 6.6.1

Consider equation (4.1) with the initial condition

$$x(t) = 0 \ (s - \eta \leq t < s, \ s \geq 0).$$

Replace in (4.1) t by $t - s$. Then (4.1) takes the form

$$\dot{x}(t - s) = \sum_{k=1}^{m} A_k x(t - s - \tau_k(t - s)) + f(t - s) \ \ (t \geq s).$$

Applying Theorem 6.4.1 to this equation, under conditions (4.2) and (4.3), we get the inequality

$$\|x(t - s)\|_{L^2(0,\infty)} \leq \|f(t - s)\|_{L^2(0,\infty)} \psi(K).$$

Or

$$\|x\|_{L^2(s,\infty)} \leq \|f\|_{L^2(s,\infty)} \psi(K). \tag{7.1}$$

Now for a fixed s, substitute

$$z(t) = W(t, s) - e^{A(t-s)} \tag{7.2}$$

into the homogeneous equation (4.5). Then

$$\dot{z}(t) = \sum_{k=1}^{m} A_k \left[W(t - \tau_k(t), s) - e^{(t-s)A}\right] = \sum_{k=1}^{m} A_k z(t - \tau_k(t)) + u(t),$$

where

$$u(t) := \sum_{k=1}^{m} A_k \left[e^{A(t-s-\tau_k(t))} - e^{A(t-s)}\right] = e^{A(t-s)} \sum_{k=1}^{m} A_k \left[e^{-A\tau_k(t)} - I\right].$$

Obviously,

$$\|u\|_{L^2(s,\infty)} \le v_0 \|e^{At}\|_{L^2(s,\infty)} \chi_A.$$

By (7.1),

$$\|z\|_{L^2(s,\infty)} \le \psi(K)\|u\|_{L^2(s,\infty)} \le \psi(K) v_0 \chi_A \|e^{At}\|_{L^2(s,\infty)}.$$

Now the estimate (6.2) follows from (7.2).

Furthermore, rewrite equation (4.5) with $w(t) = W(t,s)$ as

$$\dot{w}(t) = \sum_{k=1}^{m} A_k w(t-h_k) + \sum_{k=1}^{m} A_k [w(t-\tau_k(t)) - w(t-h_k)].$$

Then due to Lemma 6.5.1 we deduce that

$$\|\dot{w}\|_{L^2(s,\infty)} \le v_0 \|w\|_{L^2(s,\infty)} + \gamma(K)\|\dot{w}\|_{L^2(s,\infty)}.$$

Hence, condition (4.3) and the just obtained estimate for $\|W(.,s)\|_{L^2(s,\infty)}$ imply

$$\|W_t'(.,s)\|_{L^2(s,\infty)} \le \frac{v_0 \|W(.,s)\|_{L^2(s,\infty)}}{1-\gamma(K)} \le \frac{v_0}{1-\gamma(K)} (1 + \psi(K)\chi_A v_0) \|e^{At}\|_{L^2(s,\infty)}.$$

That is, inequality (6.3) is also proved.

Moreover, by the Schwarz inequality,

$$\|f(t)\|_n^2 = -\int_t^\infty \frac{d\|f(\tau)\|_n^2}{d\tau} d\tau = \int_t^\infty 2\|f(t)\|_n \|\dot{f}(\tau)\|_n d\tau$$
$$\le 2\|f\|_{L^2(s,\infty)} \|\dot{f}\|_{L^2(s,\infty)} \quad (f, \dot{f} \in L^2(s,\infty), t \ge s).$$

Hence,

$$\|W(.,s)\|_{C(s,\infty)}^2 \le 2\|W_t'(.,s)\|_{L^2(s,\infty)} \|W(.,s)\|_{L^2(s,\infty)}.$$

Now (6.2) and (6.3) imply (6.4). This proves the theorem. □

6.8 Comments

The results which appear in Section 6.1 are probably new. The material of Sections 6.2–6.4 is taken from the papers [39, 41]. Sections 6.6. and 6.7 are based on the paper [42].

Chapter 7

Periodic Systems

This chapter deals with a class of periodic systems. Explicit stability conditions are derived. The main tool is the invertibility conditions for infinite block matrices. In the case of scalar equations we apply regularized determinants.

7.1 Preliminary results

Our main object in this chapter is the equation

$$\dot{x}(t) + \sum_{s=1}^{m} A_s(t) \int_0^{\eta} x(t-\tau)d\mu_s(\tau) = 0 \ \ (0 < \eta < \infty;\ t \geq 0), \tag{1.1}$$

where μ_s are nondecreasing functions of bounded variations $\mathrm{var}(\mu_s)$, and $A_s(t)$ are T-periodic $n \times n$-matrix-valued functions satisfying conditions stipulated below.

Recall that the Floquet theory for ordinary differential equations has been developed to apply to linear periodic functional differential equations with delay, in particular, in the book [71].

Let equation (1.1) have a non-zero solution of the form

$$x(t) = p(t)e^{\lambda t}, \ \ p(t) = p(t+T).$$

Then $\mu = e^{\lambda T}$ is called the characteristic multiplier of equation (1.1) (see [71, Lemma 8.1.2]). As it was pointed in [71], a complete Floquet theory for functional differential equations is impossible. However, it is possible to define characteristic multipliers and exploit the compactness of the solution operator to show that a Floquet representation exists on the generalized eigenspace of a characteristic multiplier. The characteristic multipliers of equation (1.1) are independent of the starting time.

Lemma 7.1.1. *Equation* (1.1) *is asymptotically stable if and only if all the characteristic multipliers of equation* (1.1) *have moduli less than* 1.

For the proof see [71, Corollary 8.1.1].

Without loss of generality take $T = 2\pi$: $A_s(t) = A_s(t + 2\pi)$ $(t \in \mathbb{R}\ s = 1, \dots, m)$.

Any 2π-periodic vector-valued function f with the property $f \in L^2(0, 2\pi)$ can be represented by the Fourier series

$$f(t) = \sum_{k=-\infty}^{\infty} f_k e^{ikt} \quad \left(f_k = \frac{1}{2\pi} \int_0^{2\pi} f(t) e^{-ikt} dt;\ k = 0, \pm 1, \pm 2, \dots \right).$$

Introduce the Hilbert space PF of 2π-periodic functions defined on the real axis $\mathbb{R}$ with values in $\mathbb{C}^n$, and the scalar product

$$(f, u)_{PF} := \sum_{k=-\infty}^{\infty} (f_k, u_k)_{C^n} \quad (f, u \in PF),$$

where f_k, u_k are the Fourier coefficients of f and u, respectively. The norm in PF is

$$\|f\|_{PF} = \sqrt{(f, f)_{PF}} = \left(\sum_{k=-\infty}^{\infty} \|f_k\|_n^2 \right)^{1/2}.$$

The Parseval equality yields

$$\|f\|_{PF} = \left[\frac{1}{2\pi} \int_a^{a+2\pi} \|f(t)\|_n^2 dt \right]^{1/2} \quad (f \in PF;\ a \in \mathbb{R}).$$

In addition, introduce the subspace $DPF \subset PF$ of 2π-periodic functions f whose Fourier coefficients satisfy the condition

$$\sum_{k=-\infty}^{\infty} \|k f_k\|_n^2 < \infty.$$

That is, $f' \in L^2(0, 2\pi)$.

Furthermore, with a $\lambda \in \mathbb{C}$ substitute

$$x(t) = e^{\lambda t} v(t) \tag{1.2}$$

into (1.1). Then we have

$$\dot{v}(t) + \lambda v(t) + \sum_{s=1}^{m} A_s(t) \int_0^{\eta} e^{-\lambda \tau} v(t - \tau) d\mu_s(\tau) = 0. \tag{1.3}$$

Impose the condition

$$v(t) = v(t + 2\pi). \tag{1.4}$$

Let v_k and A_{sk} $(k = 0, \pm 1, \dots)$ be the Fourier coefficients of $v(t)$ and $A_s(t)$, respectively:

$$v(t) = \sum_{k=-\infty}^{\infty} v_k e^{ikt} \quad \text{and} \quad A_s(t) = \sum_{k=-\infty}^{\infty} A_{sk} e^{ikt} \ (s = 1, \dots, m). \tag{1.5}$$

It is assumed that

$$\sum_{k=-\infty}^{\infty} \|A_{sk}\|_n < \infty. \tag{1.6}$$

For instance, assume that $A_s(t), s = 1, \dots, m$, have integrable second derivatives from $L^2(0, 2\pi)$. Then

$$\sum_{k=-\infty}^{\infty} k^4 \|A_{sk}\|_n^2 < \infty$$

and therefore, $k^2 \|A_{sk}\|_n \to 0$ as $k \to \infty$. So condition (1.6) is fulfilled.

7.2 The main result

Without loss of generality assume that

$$\text{var}(\mu_s) = 1, \quad s = 1, \dots, m. \tag{2.1}$$

Put

$$w_0 := \sum_{s=1}^{m} \sum_{l \neq 0; -\infty}^{\infty} \|A_{sl}\|_n \quad \text{and} \quad \text{var}(F) = \sum_{s=1}^{m} \|A_{s0}\|_n.$$

Recall that the characteristic values of a matrix-valued function are the zeros of its determinant.

Now we are in a position to formulate the main result of the present chapter.

Theorem 7.2.1. *Let conditions* (1.6) *and* (2.1) *hold. Let all the characteristic values of the matrix function*

$$F(z) := zI + \sum_{s=1}^{m} A_{s0} \int_0^{\eta} e^{-z\tau} d\mu_s(\tau)$$

be in the open left half-plane C_-, and

$$w_0 \sup_{-2\,\text{var}(F) \leq \omega \leq 2\,\text{var}(F)} \|F^{-1}(i\omega)\|_n < 1. \tag{2.2}$$

Then equation (1.1) *is asymptotically stable.*

Proof. Substituting (1.5) into equation (1.3), we obtain

$$\sum_{j=-\infty}^{\infty}(ijI+\lambda)v_j e^{ijt}+\sum_{k=-\infty}^{\infty}\sum_{r=-\infty}^{\infty}\sum_{s=1}^{m}e^{irt}A_{sr}v_k\int_0^{\eta}e^{-\lambda\tau}e^{ik(t-\tau)}d\mu_s(\tau)=0, \quad (2.3)$$

or

$$(ijI+\lambda)v_j+\sum_{k=-\infty}^{\infty}\sum_{s=1}^{m}A_{s,j-k}v_k\int_0^{\eta}e^{-(\lambda+ik)\tau}d\mu_s(\tau)=0 \ \ (j=0,\pm 1,\dots).$$

Rewrite this system as

$$T(\lambda)\hat{v}=0 \ \ (\hat{v}=(v_k)_{k=-\infty}^{\infty})),$$

where $T(\lambda)=(T_{jk}(\lambda))_{j,k=-\infty}^{\infty}$ is the infinite block matrix with the blocks

$$T_{jk}(\lambda)=\sum_{s=1}^{m}A_{s,j-k}\int_0^{\eta}e^{-(\lambda+ik)\tau}d\mu_s \ \ (k\neq j)$$

and

$$T_{jj}(\lambda)=ijI+\lambda+\sum_{s=1}^{m}A_{s,0}\int_0^{\eta}e^{-(\lambda+ij)\tau}d\mu_s(\tau).$$

By [37, Theorem 6.2] (see also Section B.6 of Appendix B below) $T(z)$ is invertible, provided

$$\sup_j \|T_{jj}^{-1}(z)\|_n\sum_{\substack{k=-\infty\\k\neq j}}^{\infty}\|T_{jk}(z)\|_n<1.$$

Clearly,

$$\sum_{\substack{k=-\infty\\k\neq j}}^{\infty}\|T_{jk}(i\omega)\|_n\leq\sum_{s=1}^{m}\sum_{\substack{k=-\infty\\k\neq j}}^{\infty}\|A_{s,j-k}\|_n=\sum_{s=1}^{m}\sum_{\substack{l=-\infty\\l\neq 0}}^{\infty}\|A_{s,l}\|_n=w_0$$

and

$$\begin{aligned}&\sup_{j=1,2,\dots;\ -\infty<\omega<\infty}\|T_j^{-1}(i\omega)\|_n\\ &=\sup_{j=0,\pm1,\pm2,\dots;\ -\infty<\omega<\infty}\left\|\left(i(j+\omega)I+\sum_{s=1}^{m}A_{s,0}\int_0^{\eta}e^{-(\lambda+ij)\tau}d\mu_s(\tau)\right)^{-1}\right\|_n\\ &=\sup_{y\in(-\infty,\infty)}\left\|\left(iyI+\sum_{s=1}^{m}A_{s,0}\int_0^{\eta}e^{-iy\tau}d\mu_s(\tau)\right)^{-1}\right\|_n=\sup_{y\in(-\infty,\infty)}\|F^{-1}(iy)\|_n.\end{aligned}$$

But thanks to Lemma 4.3.1, the equality

$$\sup_{\omega\in(-\infty,\infty)} \|F^{-1}(i\omega)\|_n = \sup_{|\omega|\le 2\,\mathrm{var}(F)} \|F^{-1}(i\omega)\|_n$$

is valid.

Furthermore, for a $\xi \in (0,1]$, let us introduce the matrix $T(\xi,z) = (T_{jk}(\xi,z))$ with

$$T_{jk}(\xi,z) = \xi T_{jk}(z) \ \ (k \ne j), T_{jj}(\xi,z) = T_{jj}(z).$$

Assume that $T(z)$ has a characteristic value in the closed right-hand plane. Then according to continuity of characteristic values, for some $\xi_0 \in (0,1]$, matrix $T(\xi_0,z)$ has a characteristic value on the imaginary axis, but according to (2.2) this is impossible. This and Lemma 7.1.1 prove the theorem. □

7.3 Norm estimates for block matrices

Let A be an $n \times n$-matrix. Recall that

$$g(A) = [N_2^2(A) - \sum_{k=1}^{n} |\lambda_k(A)|^2 \,]^{1/2},$$

where $\lambda_k(A), k = 1,\dots,n$ are the eigenvalues of A, counted with their multiplicities; $N_2(A)$ is the Hilbert–Schmidt norm of A (see Section 2.3). Besides

$$g^2(A) \le N_2^2(A) - |\,\mathrm{Trace}\, A^2| \quad \text{and} \quad g^2(A) \le \frac{N_2^2(A - A^*)}{2} = 2N_2^2(A_I), \qquad (3.1)$$

where $A_I = (A - A^*)/2i$. Moreover,

$$g(e^{it}A + zI) = g(A) \ \ (t \in (-\infty,\infty); z \in \mathbb{C}). \qquad (3.2)$$

If A_1 and A_2 are commuting matrices, then $g(A_1 + A_2) \le g(A_1) + g(A_2)$. If A is a normal matrix: $AA^* = A^*A$, then $g(A) = 0$.

Let

$$B(z) = \sum_{s=1}^{m} A_{s0} \int_0^{\eta} e^{-z\tau} d\mu_s(\tau),$$

and $d(F(z))$ be the smallest modulus of eigenvalues of $F(z)$:

$$d(F(z)) = \min_{j=1,\dots,n} |\lambda_j(F(z))|.$$

Thanks to Corollary 4.3.2, the inequality

$$\sup_{|\omega|\le 2\,\mathrm{var}(F)} \|F^{-1}(i\omega)\|_n \le \Gamma_0(F)$$

is valid, where

$$\Gamma_0(F) := \sup_{|\omega|\leq 2\,\mathrm{var}(F)} \sum_{k=0}^{n-1} \frac{g^k(B(i\omega))}{\sqrt{k!}d^{k+1}(F(i\omega))}.$$

Now Theorem 7.2.1 implies our next result.

Corollary 7.3.1. *Let all the characteristic values of $F(z)$ be in C_- and $w_0\Gamma_0(F)<1$. Then equation* (1.1) *is asymptotically stable.*

Note that

$$g(B(i\omega)) \leq \sum_{s=1}^{m} N_2(A_{s0}) \quad (\omega \in \mathbb{R}).$$

If A_{s0} are mutually commutative, then

$$g(B(i\omega)) \leq \sum_{s=1}^{m} g(A_{s0}) \quad (\omega \in \mathbb{R})$$

and one can enumerate the eigenvalues of A_{s0} in such a way that

$$\lambda_j(F(z)) = z + \sum_{s=1}^{m} \lambda_j(A_{s0}) \int_0^\eta e^{-z\tau} d\mu_s(\tau).$$

Moreover, if

$$B(z) = \sum_{s=1}^{m} A_{s0}e^{-zh_s},$$

then by (3.1),

$$g(B(i\omega)) \leq \sqrt{1/2} \sum_{s=1}^{m} N_2(e^{ih_s}A_{s0} - e^{-ih_s}A_{s0}^*) \quad (\omega \in \mathbb{R}).$$

In the next section, under some assumptions, we suggest an additional simple estimate for $\Gamma_0(F)$.

7.4 Equations with one distributed delay

To illustrate Theorem 7.2.1, consider the equation

$$\dot{x}(t) + A(t) \int_0^\eta x(t-\tau)d\mu \quad (t \geq 0), \tag{4.1}$$

where $A(t)$ is a piece-wise continuous $n \times n$-matrix function, satisfying $A(t) = A(t+2\pi)$, and μ is a nondecreasing function of bounded variation.

We need the function

$$k(z) = z + \int_0^\eta \exp(-zs)d\mu(s) \ (z \in \mathbb{C}).$$

As it was shown in Section 4.6, the equality

$$\inf_{-2\,\mathrm{var}(\mu)\le\omega\le 2\,\mathrm{var}(\mu)} |k(i\omega)| = \inf_{\omega\in(-\infty,\infty)} |k(i\omega)|$$

is valid. Moreover, if

$$\mathrm{var}(\mu)\,\eta < \frac{\pi}{4}, \tag{4.2}$$

then all the zeros of $k(z)$ are in C_- and

$$\inf_{\omega\in(-\infty,\infty)} |k(i\omega)| \ge \hat{d} > 0, \tag{4.3}$$

where

$$\hat{d} := \int_0^\eta \cos(2\,\mathrm{var}(\mu)\tau)d\mu(\tau).$$

Now let A_k $(k = 0, \pm 1, \dots)$ be the Fourier coefficients of $A(t)$. Without loss of generality assume that

$$\mathrm{var}(\mu) = 1. \tag{4.4}$$

Then

$$F(z) = zI + A_0 \int_0^\eta e^{-z\tau} x(t-\tau)d\mu,$$

$$w_0 = \sum_{\substack{k=-\infty \\ k\ne 0}}^{\infty} \|A_k\|_n \quad \text{and} \quad \mathrm{var}(F) = \sum_{k=\infty}^{\infty} \|A_k\|_n.$$

According to (3.2)

$$g(B(i\omega)) = g(A_0) \ \ (\omega \in \mathbb{R})$$

and

$$\lambda_j(F(z)) = z + \lambda_j(A_0) \int_0^\eta \exp(-zs)d\mu(s) \ (z \in \mathbb{C}).$$

Let all the eigenvalues of A be real and positive:

$$0 < \lambda_1 \le \dots \le \lambda_n \tag{4.5}$$

and the conditions (4.4) and

$$\eta\,\lambda_n(A_0)\,\mathrm{var}(\mu) = \eta\,\lambda_n(A_0) < \pi/4 \tag{4.6}$$

hold, then due to (4.3),

$$\inf_{\omega\in(-\infty,\infty)} |\lambda_j(F(i\omega))| \ge \hat{d}(F),$$

where

$$\hat{d}(F) := \int_0^\eta \cos(2\lambda_n(A_0)d\mu(\tau) > 0.$$

Thus

$$\Gamma_0(F) \leq \hat{\Gamma}_0 \quad \text{where} \quad \hat{\Gamma}_0 := \sum_{k=0}^{n-1} \frac{g^k(A_0)}{\sqrt{k!}\hat{d}^{k+1}(F)}.$$

Now Corollary 7.3.1 implies the following result.

Corollary 7.4.1. *Let the conditions* (4.5), (4.6) *and*

$$w_0 \sum_{k=1}^{n-1} \frac{g^k(A_0)}{\sqrt{k!}\cos^{k+1}(2\eta\,\lambda_n(A_0))} < 1$$

hold. Then equation (4.1) *is asymptotically stable.*

If $A_0 = A_0^*$ and condition (4.5) holds, then $g(A_0) = 0$ and the stability conditions are (4.6) and

$$w_0 < \cos(2\eta\,\lambda_n(A_0)).$$

7.5 Applications of regularized determinants

In this section we are going to show that regularised determinants can be useful to investigate periodic equations. For simplicity we restrict ourselves by the simple scalar equation

$$\dot{x}(t) + bx(t) + a(t)x(t-h) = 0 \;\; (0 < h < \infty;\; t \geq 0), \tag{5.1}$$

where $a(t)$ is a real 2π-periodic piece-wise continuous scalar function and $b \neq 0$ is a real constant.

In the scalar case PF is the Hilbert space of 2π-periodic scalar functions defined on the real axis with the scalar product

$$(f,u)_{PF} := \sum_{k=-\infty}^{\infty} f_k \overline{u}_k \quad (f,u \in PF),$$

where f_k, u_k are the Fourier coefficients of f and u, respectively. The norm in PF is

$$\|f\|_{PF} = \left(\sum_{k=-\infty}^{\infty} |f_k|^2\right)^{1/2}.$$

The Parseval equality yields

$$\|f\|_{PF} = \left[\frac{1}{2\pi}\int_a^{a+2\pi} |f(t)|_n^2 dt\right]^{1/2} \quad (f \in PF;\; a \in \mathbb{R}).$$

In addition, DPF is the subspace of 2π-periodic functions f whose Fourier coefficients satisfy the condition

$$\sum_{k=-\infty}^{\infty} |kf_k|^2 < \infty.$$

Substituting

$$x(t) = e^{\lambda t} v(t) \tag{5.2}$$

into (5.1), we have

$$\dot{v}(t) + \lambda v(t) + bv(t) + a(t)e^{h\lambda} v(t-h) = 0. \tag{5.3}$$

Besides

$$v(t) = v(t + 2\pi). \tag{5.4}$$

Let v_k and a_k, $k = 0, \pm 1, \ldots$ be the Fourier coefficients of $v(t)$ and $a(t)$, respectively:

$$v(t) = \sum_{k=-\infty}^{\infty} v_k e^{ikt} \quad \text{and} \quad a(t) = \sum_{k=-\infty}^{\infty} a_k e^{ikt} . \tag{5.5}$$

Since $a(t)$ is real piece-wise continuous, we have

$$\sum_{j=-\infty}^{\infty} a_j^2 < \infty. \tag{5.6}$$

Substituting (5.5) into equation (5.3), we obtain

$$(ij + \lambda + b)v_j + \sum_{k=-\infty}^{\infty} a_{j-k} e^{-(\lambda+ik)h} v_k = 0.$$

Hence,

$$v_j + \frac{1}{ij + \lambda + b} \sum_{k=-\infty}^{\infty} a_{j-k} e^{-(\lambda+ik)h} v_k = 0 \ \ (j = 0, \pm 1, \ldots).$$

Rewrite this system as

$$(I + Z(\lambda))\hat{v} = 0 \ \ (\hat{v} = (v_k)_{k=-\infty}^{\infty}), \tag{5.7}$$

where $Z(\lambda) = (Z_{jk}(\lambda))_{j,k=-\infty}^{\infty}$ is the infinite matrix with the entries

$$Z_{jk}(\lambda) = \frac{a_{j-k} e^{-(\lambda+ik)h}}{ij + \lambda + b}.$$

For a fixed real ω we have

$$N_2^2(Z(i\omega)) = \sum_{j=-\infty}^{\infty}\sum_{k=-\infty}^{\infty} |Z_{jk}(i\omega)|^2 = \sum_{j=-\infty}^{\infty}\sum_{k=-\infty}^{\infty} \frac{|a_{j-k}e^{-(i\omega+ik)h)}|^2}{|i(j+\omega)+b|^2} \le \nu^2(\omega),$$

where

$$\nu^2(\omega) := \sum_{j=-\infty}^{\infty}\sum_{k=-\infty}^{\infty} \frac{|a_{j-k}|^2}{(j+\omega)^2+b^2}.$$

This series converges according to (5.6). Let $\lambda_k(z)$ be the eigenvalues of $Z(z)$. Then

$$\det_2(I+Z(z)) = \prod_{k=1}^{\infty}(1+\lambda_k(z))e^{-\lambda_k(z)}.$$

As it is shown in Section 1.10,

$$|\det_2(I+Z(i\omega))| \le \exp\left[\frac{\nu^2(\omega)}{2}\right].$$

We thus have established the following result.

Lemma 7.5.1. *Let the conditions* (5.6) *and* $|\det_2(I+Z(i\omega))| \ne 0$ *hold for all real* ω. *Then the problem* (5.3), (5.4) *does not have characteristic values on the real axis.*

Let us explore perturbations of (5.1). To this end consider the equation

$$\dot{x}(t) + \tilde{b}x(t) + \tilde{a}(t)x(t-\tilde{h}) \quad (0 < \tilde{h} < \infty;\ t \ge 0), \tag{5.8}$$

where $\tilde{a}(t)$ is a real 2π-periodic piece-wise continuous scalar function and $\tilde{b} \ne 0$ is a real constant.

Let $\tilde{a}_k$ $k = 0, \pm 1, \dots$ be the Fourier coefficients of $\tilde{a}(t)$. We have

$$\sum_{j=-\infty}^{\infty} \tilde{a}_j^2 < \infty. \tag{5.9}$$

In this case equation (5.4) takes the form

$$(I + \tilde{Z}(\lambda))\hat{v} = 0,$$

where $\tilde{Z}(\lambda) = (\tilde{Z}_{jk}(\lambda))_{j,k=-\infty}^{\infty}$ is an infinite matrix with the entries

$$\tilde{Z}_{jk}(\lambda) = \frac{\tilde{a}_{j-k}e^{-(\lambda+ik)\tilde{h}}}{ij+\lambda+\tilde{b}}.$$

Then

$$N_2^2(\tilde{Z}(i\omega)) = \sum_{j=-\infty}^{\infty}\sum_{k=-\infty}^{\infty} |(ij+i\omega+\tilde{b})^{-1}\tilde{a}_{j-k}e^{-(i\omega+ik)\tilde{h})}|^2 \le \tilde{\nu}^2(\omega),$$

where

$$\tilde{\nu}^2(\omega) := \sum_{j=-\infty}^{\infty} \sum_{k=-\infty}^{\infty} \frac{|\tilde{a}_{j-k}|^2}{(j+\omega)^2 + \tilde{b}^2}.$$

Besides,

$$|\det_2(I + \tilde{Z}(i\omega))| \le \exp\left[\frac{1}{2}\tilde{\nu}^2(\omega)\right].$$

Put

$$q(\omega) = N_2(Z(i\omega) - \tilde{Z}(i\omega)).$$

Then by Corollary 1.11.2 we arrive at the inequality

$$\left|\det_2 (I + Z(\omega)) - \det_2 (I + \tilde{Z}(\omega))\right| \le \delta_2(\omega),$$

where

$$\delta_2(\omega) := q(\omega) \exp\left[(1 + \nu(\omega) + \tilde{\nu}(\omega))^2/2\right].$$

Hence,

$$\left|\det_2 (I + \tilde{Z}(\omega))\right| \ge \left|\det_2 (I + Z(\omega))\right| - \delta_2(\omega).$$

So we have proved the following result.

Corollary 7.5.2. *If equation* (5.1) *is asymptotically stable and*

$$|\det_2 (I + Z(\omega))| > \delta_2(\omega) \quad (\omega \in \mathbb{R}),$$

then equation (5.8) *is also asymptotically stable.*

7.6 Comments

The material of this chapter is based on the paper [62].

As a specific case, the problem of stability investigation of linear periodic systems (LPS) with time delay is of great theoretical and practical interest. The majority of mathematical works in this area are based on investigation of the monodromy operator [69], and are mainly of theoretical nature. An application of the monodromy operator method is based on a solution of special boundary problems for ordinary differential equations and is connected with serious technical difficulties. In connection with that method, approximate approaches are used, which exploit various kinds of averaging, approximation and discretization, as well as truncation of infinite Hill determinants, [11, 75]. In the interesting paper [84], devoted to a single-loop linear periodic system with a time delay, a new approach is suggested. Namely, using the theory of the second kind integral Fredholm equations, the authors construct a characteristic function whose roots are inverses to the multipliers of the considered system. Besides, sufficient stability conditions are given, based on approximate representation of the characteristic function in the form of a polynomial.

In the present chapter we describe an alternative approach to the stability problem for a multivariable LPS with distributed delay, which is based on recent results for infinite block matrices and regularized determinants.

Chapter 8

Linear Equations with Oscillating Coefficients

In the present chapter we investigate vector and scalar linear equations with "quickly" oscillating coefficients.

8.1 Vector equations with oscillating coefficients

The present section is devoted to the following equation in $\mathbb{C}^n$:

$$\dot{x}(t) = A(t) \int_0^\eta dR_0(s)x(t-s) \quad (t \geq 0) \tag{1.1}$$

where $A(t)$ is a variable piece-wise continuous $n \times n$-matrix bounded on $[0, \infty)$; $R_0(\tau) = (r_{jk}(\tau))_{j,k=1}^n$ is an $n \times n$-matrix-valued function defined on a finite segment $[0, \eta]$, whose entries have bounded variations.

In this section we do not require that the characteristic determinant

$$\det \left(zI - A(t) \int_0^\eta e^{-zs} dR_0(s) \right)$$

is stable for all $t \geq 0$. That is, it can have zeros in the open right half-plane for some $t \geq 0$. Besides, *it is assumed that*

$$A(t) = B + C(t),$$

where B is a constant matrix such that the equation

$$\dot{y}(t) = B \int_0^\eta dR_0(s)y(t-s) \tag{1.2}$$

is exponentially stable, and $C(t)$ is a variable matrix, satisfying the condition

$$w_C := \sup_{t \geq 0} \left\| \int_0^t C(\tau)d\tau \right\|_n < \infty.$$

Recall that $\|A\|_n$ is the spectral norm of an $n \times n$ matrix A. In addition, in this section and in the next one $L^1(0,\infty) = L^1([0,\infty);\mathbb{C}^n)$, for a real nunber a and a vector function f defined and bounded on $[a,\infty)$ (not necessarily continuous) we put $\|f\|_{C(a,\infty)} = \sup_{t\geq a} \|f(t)\|_n$. Similarly, $\|A\|_{C(0,\infty)} = \sup_{t\geq 0} \|A(t)\|_n$.

Denote by $F_B(t)$ the fundamental solution to (1.2). It is not hard to check that

$$F_B(t) = \frac{1}{2\pi} \int_{-\infty}^{\infty} e^{iyt} \hat{K}_B^{-1}(iy)dy,$$

where

$$\hat{K}_B(z) = zI - B \int_0^{\eta} e^{-zs} dR_0(s).$$

Put

$$(E_0 f)(t) := \int_0^{\eta} dR_0(s) f(t-s).$$

From Lemma 1.12.1 it follows that there is a constant

$$\hat{v}(R_0) \leq \max\{\sqrt{n}\,\mathrm{var}(R_0), \sum_{j=1}^{n} \sqrt{\sum_{k=1}^{n} (\mathrm{var}(r_{jk}))^2}\},$$

such that

$$\|E_0 f\|_{C(0,\infty)} \leq \hat{v}(R_0) \|f\|_{C(-\eta,\infty)}$$

and

$$\|E_0 f\|_{L^1(0,\infty)} \leq \hat{v}(R_0) \|f\|_{L^1(-\eta,\infty)}.$$

Now we are in a position to formulate the main result of this chapter.

Theorem 8.1.1. *Assume that*

$$w_C < \frac{1}{\hat{v}\,(R_0)[1 + \hat{v}\,(R_0) \|F_B\|_{L^1(0,\infty)} (\|B\|_n + \|A\|_{C(0,\infty)})]}. \tag{1.3}$$

Then equation (1.1) *is exponentially stable.*

This theorem is proved in the next section. It is sharp. Namely, if $A(t) \equiv B$, then $w_C = 0$ and condition (1.3) is automatically fulfilled. About the estimates for $\|F_B\|_{L^1(0,\infty)}$ see for instance Section 4.4. Furthermore, let

$$A(t) = B + C_0(\omega t) \quad (\omega > 0) \tag{1.4}$$

with a piece-wise continuous matrix $C_0(t)$, such that

$$\nu_0 := \sup_{t\geq 0} \left\| \int_0^t C_0(s)ds \right\|_n < \infty. \tag{1.5}$$

Then we have

$$\|A\|_{C(0,\infty)} \leq \|B\|_n + \|C_0\|_{C(0,\infty)}$$

and

$$w_C = \sup_t \left\| \int_0^t C_0(\omega s)ds \right\|_n = \frac{\nu_0}{\omega}. \tag{1.6}$$

For example, if $C_0(t) = \sin(t)\, C_1$, with a constant matrix C_1, then $\nu_0 = 2\|C_1\|_n$. Theorem 8.1.1 implies our next result.

Corollary 8.1.2. *Assume that the conditions* (1.4), (1.5) *and*

$$\omega > \nu_0 \hat{v}\,(R_0)\left[1 + \hat{v}\,(R_0)\|F_B\|_{L^1(0,\infty)}(2\|B\|_n + \|C_0\|_{C(0,\infty)})\right] \tag{1.7}$$

hold. Then equation (1.1) *is exponentially stable.*

To illustrate Theorem 8.1.1, consider the system

$$\dot{x}(t) = A(t)\int_0^\eta x(t-\tau)d\mu(\tau), \tag{1.8}$$

where μ is a scalar nondecreasing function with $\mathrm{var}(\mu) < \infty$. Consider the equation

$$\dot{x}(t) = B\int_0^\eta x(t-\tau)d\mu(\tau), \tag{1.9}$$

assuming that B is a negative definite Hermitian $n \times n$-matrix. Again $C(t) = A(t) - B$. Put

$$E_\mu w(t) = \int_0^\eta w(t-\tau)d\mu(\tau)$$

for a scalar function $w(t)$. Reduce equation (1.9) to the diagonal form:

$$\dot{x}_j(t) = \lambda_j(B)E_\mu x_j(t) \ \ (j = 1, \dots, n), \tag{1.10}$$

where $\lambda_j(B)$ are the eigenvalues of B with their multiplicities. Let $X_j(t)$ be the fundamental solution of the scalar equation (1.10). Assume that

$$J_j := \int_0^\infty |X_j(t)|dt < \infty \ \ (j = 1, \dots, n). \tag{1.11}$$

Then the fundamental solution $\hat{F}(t)$ to (1.9) satisfies the inequality

$$\|\hat{F}\|_{L^1(0,\infty)} = \max_{j=1,\dots,n} J_j.$$

Moreover,

$$\|E_0 f\|_{C(0,\infty)} \le \mathrm{var}\,(\mu)\|f\|_{C(-\eta,\infty)}$$

and

$$\begin{aligned}\|E_0 f\|_{L^1(0,\infty)} &= \int_0^\infty \left\| \int_0^\eta f(t-\tau)d\mu(\tau)\right\|_n dt \\ &\le \int_0^\eta \int_0^\infty \|f(t-\tau)\|_n dt\, d\mu(\tau) \\ &\le \mathrm{var}(\mu)\|f\|_{L^1(-\eta,\infty)}.\end{aligned}$$

Now Theorem 8.1.1 implies

Theorem 8.1.3. *Let B be a negative definite Hermitian matrix and the conditions* (1.11) *and*

$$w_C < \frac{1}{\mathrm{var}(\mu)(1+\mathrm{var}(\mu)J_j(\|B\|_n + \|A\|_{C(0,\infty)}))} \quad (j=1,\dots,n)$$

hold. Then equation (1.8) *is exponentially stable.*

If $A(t)$ has the form (1.4) and condition (1.11) holds, then, clearly, Corollary 8.1.2 is valid with $R_0(.) = \mu(.)I$ and $\|F\|_{L^1(0,\infty)} = \max_j J_j$.

Furthermore, assume that

$$\max_{k=1,\dots,n} |\lambda_k(B)| < \frac{1}{e\ \mathrm{var}(\mu)\eta}, \tag{1.12}$$

and put

$$\rho_0(B) := \min_{k=1,\dots,n} |\lambda_k(B)|.$$

Then by Lemma 4.6.5 we have

$$J_j \le \frac{1}{\mathrm{var}(\mu)\rho_0(B)} \quad (1 \le j \le n).$$

Now Theorem 8.1.3 implies

Corollary 8.1.4. *Let B be negative definite Hermitian matrix and the conditions* (1.12), *and*

$$w_C < \frac{\rho_0(B)}{\mathrm{var}(\mu)(\rho_0(B) + \|B\|_n + \|A\|_{C(0,\infty)})}$$

hold. Then equation (1.8) *is exponentially stable.*

Now consider the equation

$$\dot{x}(t) = (B + C_0(\omega t)) \int_0^\eta x(t-\tau)d\mu(\tau). \tag{1.13}$$

So $A(t)$ has the form (1.4). Then the previous corollary and (1.6) imply

Corollary 8.1.5. *Assume that* (1.4) *and* (1.5) *hold and*

$$\rho_0(B)\omega > \nu_0\ \mathrm{var}(\mu)(\rho_0(B) + 2\|B\|_n + \|C_0\|_{C(0,\infty)}).$$

Then equation (1.13) *is exponentially stable.*

8.2 Proof of Theorem 8.1.1

For simplicity put $F_B(t) = F(t)$ and recall that

$$(E_0 f)(t) = \int_0^\eta dR_0(s) f(t-s).$$

So (1.2) can be written as

$$\dot{y}(t) = B(E_0 y)(t), \ t \geq 0.$$

Due to the Variation of Constants formula, the equation

$$\dot{x}(t) = B(E_0 x)(t) + f(t) \ \ (t \geq 0)$$

with a given function f and the zero initial condition $x(t) = 0$ $(t \leq 0)$ is equivalent to the equation

$$x(t) = \int_0^t F(t-s) f(s) ds. \tag{2.1}$$

Let $G(t,s)$ $(t \geq s \geq 0)$ be the fundamental solution to (1.1). Put $G(t,0) = G(t)$. Subtracting (1.2) from (1.1) we have

$$\frac{d}{dt}(G(t) - F(t)) \doteq B((E_0 G)(t) - (E_0 F)(t)) + C(t)(E_0 G)(t).$$

Now (2.1) implies

$$G(t) = F(t) + \int_0^t F(t-s) C(s) (E_0 G)(s) ds. \tag{2.2}$$

We need the following simple lemma.

Lemma 8.2.1. *Let $f(t), u(t)$ and $v(t)$ be matrix functions defined on a finite segment $[a,b]$ of the real axis. Assume that $f(t)$ and $v(t)$ are boundedly differentiable and $u(t)$ is integrable on $[a,b]$. Then with the notation*

$$j_u(t) = \int_a^t u(s) ds \ \ (a < t \leq b),$$

the equality

$$\int_a^t f(s) u(s) v(s) ds = f(t) j_u(t) v(t) - \int_a^t [f'(s) j_u(s) v(s) + f(s) j_u(s) v'(s)] ds$$

is valid.

Proof. Clearly,

$$\frac{d}{dt}f(t)j_u(t)v(t) = f'(t)j_u(t)v(t) + f(t)u(t)v(t) + f(t)j_u(t)v'(t).$$

Integrating this equality and taking into account that $j_u(a) = 0$, we arrive at the required result. □

Put

$$J(t) := \int_0^t C(s)ds.$$

By the previous lemma,

$$\int_0^t F(t-\tau)C(\tau)(E_0G)(\tau)d\tau$$
$$= F(0)J(t)(E_0G)(t) - \int_0^t \left[\frac{dF(t-\tau)}{d\tau}J(\tau)(E_0G)(\tau) + F(t-\tau)J(\tau)\frac{d(E_0G)(\tau)}{d\tau}\right]d\tau.$$

But $F(0) = I$,

$$\begin{aligned}\frac{d(E_0G)(\tau)}{d\tau} &= \frac{d}{d\tau}\int_0^\eta dR_0(s)G(\tau - s) = \int_0^\eta dR_0(s)\frac{d}{d\tau}G(\tau - s)\\ &= \int_0^\eta dR_0(s)\frac{d}{d\tau}G(\tau - s) = \int_0^\eta dR_0(s)A(\tau - s)(E_0G)(\tau - s)\end{aligned}$$

and

$$\frac{dF(t-\tau)}{d\tau} = -\frac{dF(t-\tau)}{dt} = B(E_0F)(t-\tau).$$

Thus,

$$\begin{aligned}&\int_0^t F(t-\tau)C(\tau)(E_0G)(\tau)d\tau\\ &= J(t)(E_0G)(t) + \int_0^t \Big[B(E_0F)(t-\tau)J(\tau)(E_0G)(\tau)\\ &\qquad - F(t-\tau)J(\tau)\int_0^\eta dR_0(s)A(\tau - s)(E_0G)(\tau - s)\Big]d\tau.\end{aligned}$$

Now (2.2) implies

Lemma 8.2.2. *The following equality is true:*

$$\begin{aligned}G(t) = F(t) + J(t)(E_0G)(t) + \int_0^t \Big[&B(E_0F)(t-\tau)J(\tau)(E_0G)(\tau)\\ &- F(t-\tau)J(\tau)\int_0^\eta dR_0(s)A(\tau - s)(E_0G)(\tau - s)\Big]d\tau.\end{aligned}$$

Take into account that

$$\|E_0 G\|_{C(0,\infty)} \le \hat{v}\ (R_0)\|G\|_{C(0,\infty)}, \|E_0 F\|_{C(0,\infty)} \le \hat{v}\ (R_0)\|F\|_{C(0,\infty)},$$

and

$$\|E_0 F\|_{L^1(0,\infty)} \le \hat{v}\ (R_0)\|F\|_{L^1(0,\infty)}.$$

Then from the previous lemma, the inequality

$$\|G\|_{C(0,\infty)} \le \|F\|_{C(0,\infty)} + \kappa\|G\|_{C(0,\infty)}$$

follows with

$$\kappa := w_C \hat{v}\ (R_0)(1 + \hat{v}\ (R_0)(\|B\|_n + \|A\|_{C(0,\infty)})\|F\|_{L^1(0,\infty)}).$$

If condition (1.3) holds, then $\kappa < 1$, and therefore,

$$\|G\|_{C(0,\infty)} = \|G(.,0)\|_{C(0,\infty)} \le \frac{\|F\|_{C(0,\infty)}}{1-\kappa}. \tag{2.3}$$

Replacing zero by s, we get the same bound for $\|G(.,s)\|_{C(s,\infty)}$. Now Corollary 3.3.1 proves the stability of (1.1). Substituting

$$x_\epsilon(t) = e^{\epsilon t} x(t) \tag{2.4}$$

with $\epsilon > 0$ into (1.1), we have the equation

$$\dot{x}_\epsilon(t) = \epsilon x_\epsilon(t) + A(t) \int_0^\eta e^{\epsilon s} dR_0(s) x_\epsilon(t-s). \tag{2.5}$$

If $\epsilon > 0$ is sufficiently small, then considering (2.5) as a perturbation of the equation

$$\dot{y}(t) = \epsilon y(t) + B \int_0^\eta e^{\epsilon s} dR_0(s) y(t-s)$$

and applying our above arguments, according to (2.3), we obtain $\|x_\epsilon\|_{C(0,\infty)} < \infty$ for any solution x_ϵ of (2.5). Hence (2.4) implies

$$\|x(t)\|_n \le e^{-\epsilon t}\|x_\epsilon\|_{C(0,\infty)} \quad (t \ge 0)$$

for any solution x of (1.1), as claimed. □

8.3 Scalar equations with several delays

In the present section, in the case of scalar equations, we particularly generalize the results of the previous section to equations with several delays. Namely, this section deals with the equation

$$\dot{x}(t) + \sum_{j=1}^{m} a_j(t) \int_0^h x(t-s) dr_j(s) = 0 \quad (t \ge 0;\ h = \text{const} > 0), \tag{3.1}$$

where $r_j(s)$ are nondecreasing functions having finite variations $\mathrm{var}(r_j)$, and $a_j(t)$ are piece-wise continuous real functions bounded on $[0, \infty)$.

In the present section *we do not require that $a_j(t)$ are positive for all $t \geq 0$. So the function*

$$z + \sum_{j=1}^{m} a_j(t) \int_0^h e^{-zs} dr_j(s)$$

can have zeros in the right-hand plane for some $t \geq 0$.

Let

$$a_j(t) = b_j + c_j(t) \quad (j = 1, \dots, m),$$

where b_j are positive constants, such that *all the zeros of the function*

$$k(z) := z + \sum_{j=1}^{m} b_j \int_0^h e^{-zs} dr_j(s)$$

are in the open left-hand plane, and functions $c_j(t)$ have the property

$$w_j := \sup_{t \geq 0} \left| \int_0^t c_j(t) dt \right| < \infty \quad (j = 1, \dots, m).$$

The function

$$W(t) = \frac{1}{2\pi i} \int_{-i\infty}^{i\infty} \frac{e^{zt} dz}{k(z)}$$

is the fundamental solution to the equation

$$\dot{y}(t) = -\sum_{j=1}^{m} b_j \int_0^h y(t-s) dr_j(s). \tag{3.2}$$

Without loss of generality assume that

$$\mathrm{var}(r_j) = 1 \quad (j = 1, \dots, m).$$

In this section and in the next one, $L^1 = L^1(0, \infty)$ is the space of real scalar functions integrable on $[0, \infty)$. So

$$\|W\|_{L^1} = \int_0^\infty |W(t)| dt.$$

For a scalar function f defined and bounded on $[0, \infty)$ (not necessarily continuous) we put $\|f\|_C = \sup_{t \geq 0} |f(t)|$.

Theorem 8.3.1. *Let*

$$\left(\sum_{j=1}^{m} w_j \right) \left[1 + \|W\|_{L^1} \sum_{k=1}^{m} (b_k + \|a_k\|_C) \right] < 1. \tag{3.3}$$

Then equation (3.1) is exponentially stable.

This theorem is proved in the next section. It is sharp. Namely, if $a_j(t) \equiv b_j$ $(j = 1, \dots, m)$, then $w_j = 0$ and condition (3.3) is automatically fulfilled.

Let

$$\hat{r}(s) = \sum_{j=1}^{m} b_j r_j(s).$$

Then (3.2) takes the form

$$\dot{y}(t) = -\int_0^h y(t-s)d\hat{r}(s). \tag{3.4}$$

Besides,

$$\operatorname{var}(\hat{r}) = \sum_{j=1}^{m} b_j \ \operatorname{var}(r_j) = \sum_{j=1}^{m} b_j.$$

For instance, let

$$eh\sum_{j=1}^{m} b_j = eh \ \operatorname{var}(\hat{r}) < 1. \tag{3.5}$$

Then $W(t) \geq 0$ and equation (3.2) is exponentially stable, cf. Section 4.6. Now, integrating (3.2), we have

$$\begin{aligned} 1 = W(0) &= \int_0^\infty \int_0^h W(t-s)d\hat{r}(s)\, dt = \int_0^h \int_0^\infty W(t-s)\, dt\, d\hat{r}(s) \\ &= \int_0^h \int_{-s}^\infty W(t)\, dt\, d\hat{r}(s) = \int_0^h \int_0^\infty W(t)\, dt\, d\hat{r}(s) = \operatorname{var}(\hat{r})\|W\|_{L^1}. \end{aligned}$$

So

$$\|W\|_{L^1} = \frac{1}{\sum_{k=1}^m b_k}. \tag{3.6}$$

Thus, Theorem 8.3.1 implies

Corollary 8.3.2. *Let the conditions* (3.5) *and*

$$\sum_{j=1}^{m} w_j < \frac{\sum_{k=1}^m b_k}{\sum_{k=1}^m (2b_k + \|a_k\|_C)} \tag{3.7}$$

hold. Then equation (3.1) *is exponentially stable.*

Furthermore, let

$$a_j(t) = b_j + u_j(\omega_j t) \ \ (\omega_j > 0;\ j = 1, \dots, m) \tag{3.8}$$

with a piece-wise continuous functions $u_j(t)$, such that

$$\nu_j := \sup_t \left| \int_0^t u_j(s)ds \right| < \infty.$$

Then we have $\|a_j\|_C \le b_j + \|u_j\|_C$ and

$$w_j = \sup_t |\int_0^t u_j(\omega_j s)ds| = \nu_j/\omega_j.$$

For example if $u_j(t) = \sin(t)$, then $\nu_j = 2$. Now Theorem 8.3.1 and (3.7) imply our next result.

Corollary 8.3.3. *Let the conditions* (3.5), (3.8) *and*

$$\sum_{k=1}^{m} \frac{\nu_j}{\omega_j} < \frac{\sum_{k=1}^m b_k}{\sum_{k=1}^m (3b_k + \|u_k\|_C)} \tag{3.9}$$

hold. Then equation (3.1) *is exponentially stable.*

Example 8.3.4. Consider the equation

$$\dot{x} = -\sum_{j=1}^{m} (b_j + \tau_j \sin(\omega_j t)) \int_0^1 x(t-s) d_j(s) ds \ \ (\tau_j = \text{const} > 0), \tag{3.10}$$

where $d_j(s)$ are positive and bounded on $[0,1]$ functions, satisfying the condition

$$\int_0^1 d_j(s)ds = 1.$$

Assume that (3.5) holds with $h = 1$. Then $\nu_j = 2\tau_j$ and condition (3.9) takes the form

$$\sum_{k=1}^{m} \frac{2\tau_j}{\omega_j} < \frac{\sum_{k=1}^m b_k}{\sum_{k=1}^m (3b_k + \tau_k)}.$$

So for arbitrary τ_j, there are ω_j, such that equation (3.10) is exponentially stable. In particular, consider the equation

$$\dot{x} = -(b + \tau_0 \sin(\omega t)) x(t-1) \ \ (b < e^{-1}; \tau_0, \omega = \text{const} > 0). \tag{3.11}$$

Then according to condition (3.9) for any τ_0, there is an ω, such that equation (3.11) is exponentially stable.

Furthermore, assume that

$$eh \sum_{j=1}^{m} b_j < \xi \ \ (1 < \xi < 2) \tag{3.12}$$

and consider the equation

$$\dot{y}(t) + \sum_{j=1}^{m} \tilde{b}_j \int_0^h y(t-s) dr_j(s) = 0, \tag{3.13}$$

where $\tilde{b}_j = b_j/\xi$. Let $\tilde{W}$ be the fundamental solution to the equation (3.13). Subtracting (3.13) from (3.2), we obtain

$$\frac{d}{dt}(W(t) - \tilde{W}(t)) + \sum_{j=1}^{m} \tilde{b}_j \int_0^h (W(t-s) - \tilde{W}(t-s))dr_j(s)$$
$$= -\sum_{j=1}^{m} (b_j - \tilde{b}_j) \int_0^h W(t-s)dr_j(s).$$

Due to the Variation of Constants formula,

$$W(t) - \tilde{W}(t) = -\int_0^t \tilde{W}(t-\tau) \sum_{j=1}^{m} (b_j - \tilde{b}_j) \int_0^h W(\tau - s)dr_j(s)d\tau.$$

Hence, taking into account that $\mathrm{var}(r_j) = 1$, by simple calculations we get

$$\|W - \tilde{W}\|_{L^1} \le \|\tilde{W}\|_{L^1} \|W\|_{L^1} \sum_{j=1}^{m} (b_j - \tilde{b}_j).$$

If

$$\psi := \|\tilde{W}\|_{L^1} \sum_{j=1}^{m} (b_j - \tilde{b}_j) < 1,$$

then

$$\|W\|_{L^1} \le \frac{\|\tilde{W}\|_{L^1}}{1 - \psi}.$$

But condition (3.12) implies (3.5) with $\tilde{b}_k$ instead of b_k. So according to (3.6) we have

$$\|\tilde{W}\|_{L^1} = \frac{1}{\sum_{k=1}^{m} \tilde{b}_k} = \frac{\xi}{\sum_{k=1}^{m} b_k}.$$

Consequently, $\psi = \xi - 1$ and

$$\|W\|_{L^1} \le \frac{\|\tilde{W}\|_{L^1}}{2 - \xi}.$$

Thus we have proved the following result.

Lemma 8.3.5. *Let conditions* (3.12) *and* $\mathrm{var}(r_j) = 1$ $(j = 1, \dots, m)$ *hold. Then*

$$\|W\|_{L^1} \le \frac{\xi}{(2 - \xi) \sum_{k=1}^{m} b_k}.$$

Now we can directly apply Theorem 8.3.1.

8.4 Proof of Theorem 8.3.1

Due to the Variation of Constants formula the equation

$$\dot{x}(t) = -\sum_{k=1}^{m} b_j \int_0^h x(t-s)dr_j(s) + f(t) \ \ (t > 0)$$

with a given function f and the zero initial condition

$$x(t) = 0 \ (t \le 0)$$

is equivalent to the equation

$$x(t) = \int_0^t W(t-\tau)f(\tau)d\tau. \tag{4.1}$$

Recall that a differentiable in t function $G(t,\tau)$ $(t \ge \tau \ge 0)$ is the fundamental solution to (3.1) if it satisfies that equation in t and the initial conditions

$$G(\tau,\tau) = 1, \ G(t,\tau) = 0 \ (t < \tau, \ \tau \ge 0).$$

Put $G(t,0) = G(t)$. Subtracting (3.2) from (3.1) we have

$$\begin{aligned}\frac{d}{dt}(G(t) - W(t)) = & -\sum_{k=1}^{m} b_j \int_0^h (G(t-s) - W(t-s))dr_j(s) \\ & -\sum_{j=1}^{m} c_j(t) \int_0^h G(t-s)dr_j(s).\end{aligned} \tag{4.2}$$

Now (4.1) implies

$$G(t) = W(t) - \int_0^t W(t-\tau) \sum_{j=1}^{m} c_j(\tau) \int_0^h G(\tau - s)dr_j(s) \, d\tau. \tag{4.3}$$

Put

$$J_j(t) := \int_0^t c_j(s)ds.$$

By Lemma 8.2.1 we obtain the inequality

$$\begin{aligned}&\int_0^t W(t-\tau)c_j(\tau)G(\tau-s)d\tau \\ &\quad = W(0)J_j(t)G(t-s) \\ &\quad - \int_0^t \left[\frac{dW(t-\tau)}{d\tau} J_j(\tau)G(\tau-s) + W(t-\tau)J_j(\tau)\frac{dG(\tau-s)}{d\tau}\right] d\tau.\end{aligned} \tag{4.4}$$

But

$$\frac{dG(\tau - s)}{d\tau} = -\sum_{k=1}^{m} a_k(\tau - s) \int_0^h G(\tau - s - s_1) dr_k(s_1)$$

and

$$\frac{dW(t-\tau)}{d\tau} = -\frac{dW(t-\tau)}{dt} = \sum_{j=1}^{m} b_j \int_0^h W(t-\tau-s_1) dr_j(s_1)$$
$$= \int_0^h W(t-\tau-s_1) d\hat{r}(s_1).$$

Thus,

$$\int_0^t W(t-\tau) c_j(\tau) G(\tau - s) d\tau = Z_j(t,s),$$

where

$$Z_j(t,s) := J_j(t)G(t-s) + \int_0^t J_j(\tau)\Big[-\int_0^h W(t-\tau-s_1)d\hat{r}(s_1)G(\tau-s)$$
$$+ W(t-\tau)\sum_{k=1}^{m} a_k(\tau-s)\int_0^h G(\tau-s-s_1)dr_k(s_1)\Big] d\tau.$$

Now (4.3) implies

Lemma 8.4.1. *The equality*

$$G(t) = W(t) - \int_0^h \sum_{j=1}^{m} Z_j(t,s) dr_j(s)$$

is true.

We have

$$\sup_{t\geq 0} |Z_j(t,s)|$$
$$\leq w_j \|G\|_C \left[1 + \int_0^t \int_0^h |W(t-\tau-s_1)| d\hat{r}(s_1) d\tau + \sum_{k=1}^{m} \|a_k\|_C \int_0^t |W(t-\tau)| d\tau\right].$$

But

$$\int_0^h \int_0^t |W(t-\tau-s)| d\tau\, d\hat{r}(s) = \int_0^h \int_{-s}^{t-s} |W(t-\tau)| d\tau\, d\hat{r}(s)$$
$$\leq \operatorname{var}(\hat{r}) \int_0^\infty |W(\tau)| d\tau.$$

Thus,

$$\|Z_j(t,s)\|_C \leq w_j \|G\|_C \left(1 + \sum_{k=1}^{m} (b_k + \|a_k\|_C) \|W\|_{L^1}\right).$$

From the previous lemma we get $\|G\|_C \leq \|W\|_C + \gamma\|G\|_C$, where

$$\gamma := \left(\sum_{k=1}^{m} w_j\right)\left[1 + \sum_{k=1}^{m}(b_k + \|a_k\|_C)\|W\|_{L^1}\right].$$

Condition (3.3) means that $\gamma < 1$.

We thus have proved the following result.

Lemma 8.4.2. *Let condition* (3.3) *hold. Then*

$$\|G\|_C \leq \frac{\|W\|_C}{1-\gamma}. \tag{4.5}$$

The previous lemma implies the stability of (3.1). Substituting

$$x_\epsilon(t) = e^{-\epsilon t}x(t) \tag{4.6}$$

with $\epsilon > 0$ into (3.1), we have the equation

$$\dot{x}_\epsilon(t) = \epsilon x_\epsilon(t) - \sum_{k=1}^{m} a_k(t)\int_0^h e^{\epsilon s}x_\epsilon(t-s)dr_k(s). \tag{4.7}$$

If $\epsilon > 0$ is sufficiently small, then according to (4.5) we easily obtain that $\|x_\epsilon\|_C < \infty$ for any solution x_ϵ of (4.7). Hence (4.6) implies

$$|x(t)| \leq e^{-\epsilon t}\|x_\epsilon\|_C \quad (t \geq 0)$$

for any solution x of (3.1), as claimed. □

8.5 Comments

This chapter is based on the papers [54] and [55].

The literature on the first-order scalar linear functional differential equations is very rich, cf. [81, 89, 102, 109, 114] and references therein, but mainly, the coefficients are assumed to be positive. The papers [6, 7, 117] are devoted to stability properties of differential equations with several (not distributed) delays and an arbitrary number of positive and negative coefficients. In particular, the papers [6, 7] give us explicit stability tests in the iterative and limit forms. Besides the main tool is the comparison method based on the Bohl–Perron type theorem. The sharp stability condition for the first-order functional-differential equation with one variable delay was established by A.D. Myshkis (the so-called 3/2-stability theorem) in his celebrated paper [94]. A similar result was established by J. Lillo [87]. The 3/2-stability theorem was generalized to nonlinear equations and equations with unbounded delays in the papers [112, 113, 114]. As Example 8.3.4 shows, Theorem 8.3.1 improves the 3/2-stability theorem in the case of constant delays and "quickly" oscillating coefficients.

It should be noted that the theory of vector functional differential equations with oscillating coefficients in contrast to scalar equations is not enough developed.

Chapter 9

Linear Equations with Slowly Varying Coefficients

This chapter deals with vector differential-delay equations having slowly varying coefficients. The main tool in this chapter is the "freezing" method.

9.1 The "freezing" method

Again consider in $\mathbb{C}^n$ the equation

$$\dot{y}(t) = \int_0^\eta d_\tau R(t,\tau) y(t-\tau) \quad (t \geq 0), \tag{1.1}$$

where $R(t,\tau) = (r_{jk}(t,\tau))_{j,k=1}^n$ is an $n \times n$- matrix-valued function defined on $[0,\infty) \times [0,\eta]$ whose entries have uniformly bounded variations in τ. $C(a,b)$ is the space of continuous vector-valued functions, again. It is also assumed that there is a positive constant q, such that

$$\left\| \int_0^\eta d_\tau (R(t,\tau) - R(s,\tau)) f(t-\tau) \right\|_n \leq q \, |t-s| \|f\|_{C(-\eta,t)} \tag{1.2}$$
$$(f \in C(-\eta,t);\ t,s \geq 0).$$

For example, consider the equation

$$\dot{y}(t) = \int_0^\eta A(t,\tau) y(t-\tau) d\tau + \sum_{k=0}^m A_k(t) y(t-h_k) \quad (t \geq 0;\ m < \infty) \tag{1.3}$$

where $0 = h_0 < h_1 < \cdots < h_m \leq \eta$ are constants, $A_k(t)$ and $A(t,\tau)$ are matrix-valued functions satisfying the inequality

$$\int_0^\eta \|A(t,\tau) - A(s,\tau)\|_n d\tau + \sum_{k=0}^m \|A_k(t) - A_k(s)\|_n \leq q \, |t-s| \ (t,s \geq 0). \tag{1.4}$$

Then we have

$$\left\| \int_0^\eta (A(t,\tau) - A(s,\tau))f(t-\tau)d\tau + \sum_{k=0}^m (A_k(t) - A_k(s))f(t-h_k) \right\|_n$$

$$\leq \|f\|_{C(-\eta,t)} \left(\int_0^\eta \|A(t,\tau) - A(s,\tau)\|_n d\tau + \sum_{k=0}^m \|A_k(t) - A_k(s,\tau)\|_n \right).$$

So condition (1.2) holds.

To formulate the result, for a fixed $s \geq 0$, consider the "frozen" equation

$$\dot{y}(t) = \int_0^\eta d_\tau R(s,\tau) y(t-\tau) \ \ (t > 0). \tag{1.5}$$

Let $G_s(t)$ be the fundamental solution to the autonomous equation (1.5).

Theorem 9.1.1. *Let the conditions* (1.2) *and*

$$\chi := \sup_{s \geq 0} \int_0^\infty t \|G_s(t)\|_n dt < \frac{1}{q} \tag{1.6}$$

hold. Then equation (1.1) *is exponentially stable.*

This theorem is proved in the next section. Let us establish an estimate for χ. Recall that $K_s(z) = zI - \int_0^\eta e^{-z\tau} d_\tau R(s,\tau)$.

Put

$$\alpha_0 := \sup_{s \geq 0;\ k=1,2,\dots} \operatorname{Re} z_k(K_s),$$

where $z_k(K_s)$ are the characteristic values of K_s; so under our assumptions $\alpha_0 < 0$. Since $(K_s^{-1}(z))' = dK_s^{-1}(z)/dz$ is the Laplace transform of $-tG_s(t)$, for any positive $c < |\alpha_0|$, we obtain

$$tG_s(t) = -\frac{1}{2\pi i} \int_{-i\infty}^{i\infty} e^{zt} (K_s^{-1}(z))' dz = -\frac{1}{2\pi i} \int_{-\infty}^{\infty} e^{t(i\omega - c)} (K_s^{-1}(i\omega - c))' d\omega.$$

Hence,

$$t\|G_s(t)\|_n \leq \frac{e^{-tc}}{2\pi} \int_{-\infty}^{\infty} \|(K_s^{-1}(i\omega - c))'\|_n d\omega.$$

But $(K_s^{-1}(z))' = -K_s^{-1}(z) K_s'(z) K_s^{-1}(z)$ and thus

$$(K_s^{-1}(i\omega - c))' = -K_s^{-1}(i\omega - c) \left(I + \int_0^\eta \tau e^{-(i\omega - c)\tau} d_\tau R(s,\tau) \right) K_s^{-1}(i\omega - c).$$

Therefore,

$$t\|G_s(t)\|_n \leq \frac{M_{c,s} e^{-tc}}{2\pi} \int_{-\infty}^{\infty} \|K_s^{-1}(i\omega - c)\|_n^2 d\omega,$$

where

$$M_{c,s} = \sup_{\omega} \left\| I + \int_0^{\eta} \tau e^{-(i\omega - c)\tau} d_\tau R(s,\tau) \right\|_n \leq 1 + e^{c\eta} \text{ vd}(R(s,.))$$
$$\leq 1 + \eta\, e^{c\eta} \text{ var}\, R(s,.).$$

Here vd$(R(s,.)$ is the spectral norm of the matrix whose entries are

$$\int_0^{\eta} \tau d_\tau |r_{jk}(s,\tau)|.$$

We thus obtain the following result.

Lemma 9.1.2. *For any positive $c < |\alpha_0|$ we have*

$$\chi \leq \sup_{s \geq 0} \frac{M_{c,s}}{2\pi c} \int_{-\infty}^{\infty} \|K_s^{-1}(i\omega - c)\|_n^2 d\omega.$$

9.2 Proof of Theorem 9.1.1

Again consider the non-homogeneous equation

$$\dot{x}(t) = \int_0^{\eta} d_\tau R(t,\tau) x(t-\tau) + f(t) \quad (t \geq 0) \tag{2.1}$$

with a given $f \in C(0,\infty)$ and the zero initial condition

$$x(t) = 0 \quad (t \leq 0). \tag{2.2}$$

For a continuous vector-valued function u defined on $[-\eta\,,\infty)$ and a fixed $s \geq 0$, put

$$(E(s)u)(t) = \int_0^{\eta} d_\tau R(s,\tau) u(t-\tau).$$

Then (2.1) can be written as

$$\dot{x}(t) = E(s)x(t) + [E(t) - E(s)]x(t) + f(t).$$

By the Variation of Constants formula

$$x(t) = \int_0^t G_s(t-t_1)[(E(t_1) - E(s))x(t_1) + f(t_1)]dt_1. \tag{2.3}$$

Condition (1.2) gives us the inequality

$$\|[E(t) - E(s)]x(t)\|_n \leq q|t-s|\|x\|_{C(0,t)} \quad (t, s \geq 0). \tag{2.4}$$

Note that for an $\epsilon > 0$, we have

$$\int_0^\infty \|G_s(t)\|_n dt \le \int_0^\epsilon \|G_s(t)\|_n dt + \frac{1}{\epsilon}\int_\epsilon^\infty t\|G_s(t)\|_n dt \le c_1$$

where

$$c_1 = \sup_{s\ge 0}\int_0^\epsilon \|G_s(t)\|_n dt + \frac{1}{\epsilon}\chi.$$

Thus

$$\left\|\int_0^t G_s(t-t_1)f(t_1)dt_1\right\|_n \le \|f\|_{C(0,\infty)}\int_0^t \|G_s(t_1)\|_n dt_1 \le c_1\|f\|_{C(0,\infty)}.$$

Now (2.4) and (2.3) for a fixed $t > 0$ imply

$$\|x(t)\|_n \le c_1\|f\|_{C(0,\infty)} + \|x\|_{C(0,t)}\int_0^t \|G_s(t-t_1)\|_n q|t_1 - s|dt_1 \quad (t \ge 0).$$

Hence with $s = t$ we obtain

$$\|x(t)\|_n \le c_1\|f\|_{C(0,\infty)} + \|x\|_{C(0,t)}\int_0^t \|\hat{G}_t(t-t_1)\|_n q(t-t_1)dt_1. \tag{2.5}$$

But

$$\int_0^t \|\hat{G}_t(t-t_1)\|_n (t-t_1)dt_1 = \int_0^t \|\hat{G}_t(u)\|_n u\, du$$
$$\le \int_0^\infty \|\hat{G}_t(u)\|_n u du \le \sup_{t\ge 0}\int_0^\infty \|\hat{G}_t(u)\|_n u\, du = \chi.$$

Thus (2.5) implies

$$\|x\|_{C(0,T)} \le c_1\|f\|_{C(0,\infty)} + \|x\|_{C(0,T)}\chi q$$

for any finite $T > 0$. Hence, due to condition (1.6) we arrive at the inequality

$$\|x\|_{C(0,T)} \le \frac{c_1\|f\|_{C(0,\infty)}}{1-\chi q}.$$

Hence, letting $T \to \infty$ we get

$$\|x\|_{C(0,\infty)} \le \frac{c_1\|f\|_{C(0,\infty)}}{1-q\chi}.$$

So for any bounded f the solution of problem (2.1), (2.2), is uniformly bounded. Now Theorem 3.4.1 proves the required result. □

9.3 Perturbations of certain ordinary differential equations

Now let us consider the equation

$$\dot{x}(t) = A(t)y(t) + \int_0^\eta d_\tau R_1(t,\tau)y(t-\tau) \ \ (t \geq 0), \tag{3.1}$$

where $A(t)$ for any $t \geq 0$ is an $n \times n$ Hurwitzian matrix satisfying the condition

$$\|A(t) - A(s)\|_n \leq q_0|t-s| \ (t, s \geq 0), \tag{3.2}$$

and $R_1(t,\tau)$ is an $n \times n$ matrix-valued function defined on $[0,\infty) \times [0,\eta\,]$ whose entries have uniformly bounded variations. Put

$$E_1 u(t) = \int_0^\eta d_\tau R_1(s,\tau)u(t-\tau) \ \ (u \in C(-\eta\,,\infty)).$$

By Lemma 1.12.3, there is a constant $V(R_1)$ such that

$$\|E_1 u\|_{C(0,\infty)} \leq V(R_1)\|u\|_{C(-\eta\,,\infty)}. \tag{3.3}$$

Besides, some estimates for $V(R_1)$ are given in Section 1.12. Let us appeal to a stability criterion which is more convenient than Theorem 9.1.1 in the case of equation (3.1).

Theorem 9.3.1. *Let the conditions* (3.2),

$$\nu_A := \sup_{s\geq 0} \int_0^\infty \|e^{A(s)t}\|_n dt \ < \frac{1}{V(R_1)} \tag{3.4}$$

and

$$\hat{\chi}_0 := \sup_{s\geq 0} \int_0^\infty t\|e^{A(s)t}\|_n dt < \frac{1-\nu_A V(R_1)}{q_0} \tag{3.5}$$

hold. Then equation (3.1) *is exponentially stable.*

This theorem is proved in the next section. For instance, consider the equation

$$\dot{x}(t) = A(t)x(t) + \sum_{k=1}^m B_k(t)\int_0^\eta x(t-\tau)d\mu_k(\tau) \ \ (t \geq 0) \tag{3.6}$$

where μ_k are nondecreasing scalar functions, and $B_k(t)$ are $n \times n$-matrices with the properties

$$\sup_{t\geq 0} \|B_k(t)\|_n < \infty \ \ (k = 1, \dots, m).$$

Simple calculations show that in the considered case

$$V(R_1) \leq \sup_s \sum_{k=1}^m \|B_k(s)\|_n \operatorname{var}(\mu_k).$$

About various estimates for $\|e^{A(s)t}\|_n$ see Sections 2.5 and 2.8.

9.4 Proof of Theorems 9.3.1

To prove Theorem 9.3.1, consider the equation

$$\dot{x}(t) = A(t)x(t) + \int_0^\eta d_\tau R_1(t,\tau)x(t-\tau) + f(t) \tag{4.1}$$

and denote by $U(t,s)$ $(t \geq s \geq 0)$ the evolution operator of the equation

$$\dot{y}(t) = A(t)y(t). \tag{4.2}$$

Put

$$\xi_A := \sup_{f \in C(0,\infty)} \frac{1}{\|f\|_{C(0,\infty)}} \sup_{t \geq 0} \left\| \int_0^t U(t,s)f(s)ds \right\|_n .$$

Lemma 9.4.1. *Let the condition*

$$\xi_A V(R_1) < 1 \tag{4.3}$$

hold. Then any solution of (4.1) *with* $f \in C(0,\infty)$ *and the zero initial condition satisfies the inequality*

$$\|x\|_{C(0,\infty)} \leq \frac{\xi_A \|f\|_{C(0,\infty)}}{1 - V(R_1)\xi_A}.$$

Proof. Equation (4.1) is equivalent to the following one:

$$x(t) = \int_0^t U(t,s)(E_1 x(s) + f(s))ds.$$

Hence,

$$\|x\|_{C(0,\infty)} \leq \xi_A(\|E_1 x\|_{C(0,\infty)} + \|f\|_{C(0,\infty)}).$$

Hence, by (3.3) we arrive at the inequality

$$\|x\|_{C(0,\infty)} \leq \xi_A(V(R_1)\|x\|_{C(0,\infty)} + \|f\|_{C(0,\infty)}).$$

Now condition (4.3) ensures the required result. □

The previous lemma and Theorem 3.4.1 imply

Corollary 9.4.2. *Let condition* (4.3) *hold. Then equation* (3.1) *is exponentially stable.*

Lemma 9.4.3. *Let conditions* (3.2) *and* (3.5) *hold. Then*

$$\xi_A < \frac{\nu_A}{1 - q_0 \hat{\chi}_0}.$$

Proof. Consider the equation

$$\dot{x}(t) = A(t)x(t) + f(t) \tag{4.4}$$

with the zero initial condition $x(0) = 0$. Rewrite it as

$$\dot{x}(t) = A(s)x(t) + (A(t) - A(s))x(t) + f(t).$$

Hence

$$x(t) = \int_0^t e^{A(s)(t-t_1)}[(A(t_1) - A(s))x(t_1) + f(t_1)]dt_1.$$

Take $s = t$. Then

$$\|x(t)\|_n \leq \int_0^t \|e^{A(t)(t-t_1)}\| \|(A(t_1) - A(t))x(t_1)\| dt_1 + c_0,$$

where

$$c_0 := \sup_{s,t} \int_0^t \|e^{A(s)(t-t_1)}\|_n \|f(t_1)\|_n dt_1$$
$$\leq \|f\|_{C(0,\infty)} \sup_s \int_0^\infty \|e^{A(s)t_1}\|_n dt_1 \leq \nu_A \|f\|_{C(0,\infty)}.$$

Thus, for any $T < \infty$, we get

$$\sup_{t\leq T} \|x(t)\|_n \leq c_0 + q_0 \sup_{t\leq T} \|x(t)\|_n \int_0^T \|e^{A(t)(T-t_1)}\| |t_1 - T| dt_1$$
$$\leq c_0 + q_0 \sup_{t\leq T} \|x(t)\|_n \int_0^T \|e^{A(t)u}\| u du$$
$$\leq c_0 + q_0 \hat{\chi}_0 \sup_{t\leq T} \|x(t)\|_n.$$

By (3.5), we have $q_0\hat{\chi}_0 < 1$. So

$$\|x\|_{C(0,T)} \leq \frac{c_0}{1 - q_0\hat{\chi}_0} = \frac{\nu_A \|f\|_{C(0,\infty)}}{1 - q_0\hat{\chi}_0}.$$

Hence, letting $T \to \infty$, we get

$$\|x\|_{C(0,\infty)} \leq \frac{c_0}{1 - q_0\hat{\chi}_0} = \frac{\nu_A \|f\|_{C(0,\infty)}}{1 - q_0\hat{\chi}_0}.$$

This proves the lemma. □

Proof of Theorem 9.3.1. The required result at once follows from Corollary 9.4.2 and the previous lemma. □

9.5 Comments

The papers [23] and [60] were essential to this chapter.

Theorem 9.1.1 extends the "freezing" method for ordinary differential equations, cf. [12, 76, 107]. Nonlinear systems with delay and slowly varying coefficient were considered in [29].

Chapter 10

Nonlinear Vector Equations

In the present chapter we investigate nonlinear systems with causal mappings. The main tool is that of norm estimates for fundamental solutions. The generalized norm is also applied. It enables us to use information about a system more completely than the usual (number) norm.

10.1 Definitions and preliminaries

Let $\eta < \infty$ be a positive constant, and $R(t,\tau) = (r_{jk}(t,\tau))_{j,k=1}^n$ be an $n \times n$-matrix-valued function defined on $[0,\infty) \times [0,\eta\,]$, piece-wise continuous in t for each τ, whose entries have uniformly bounded variations in τ:

$$v_{jk} = \sup_{t \geq 0} \operatorname{var} r_{jk}(t,.) < \infty \quad (j,k = 1,\dots,n).$$

In this chapter, again $C([a,b],\mathbb{C}^n) = C(a,b)$ and $L^p([a,b],\mathbb{C}^n) = L^p(a,b)$. Our main object in the present section is the problem

$$\dot{x}(t) = \int_0^\eta d_s R(t,s)x(t-s) + [Fx](t) + f(t) \quad (t \geq 0), \tag{1.1}$$

$$x(t) = \phi(t) \in C(-\eta\,,0) \quad (-\eta\, \leq t \leq 0), \tag{1.2}$$

where $f \in C(0,\infty)$ and F is a continuous causal mapping in $C(-\eta\,,\infty)$ (see Section 1.8).

Let

$$\Omega(\varrho) := \{v \in C(-\eta\,,\infty) : \|v\|_{C(-\eta\,,\infty)} \leq \varrho\}$$

for a positive $\varrho \leq \infty$.

It is supposed that there is a constant $q \geq 0$, such that

$$\|Fw\|_{C(0,\infty)} \leq q\|w\|_{C(-\eta\,,\infty)} \; (w \in \Omega(\varrho)). \tag{1.3}$$

Below we present some examples of the mapping satisfying condition (1.3).

Lemma 10.1.1. *Let F be a continuous causal mapping in $C(-\eta,\infty)$ and condition* (1.3) *hold. Then F is a continuous mapping in $C(-\eta,T)$ and*

$$\|Fw\|_{C(0,T)} \leq q\|w\|_{C(-\eta,T)} \quad (w \in \Omega(\varrho) \cap C(-\eta,T))$$

for all $T > 0$.

Proof. Take $w \in \Omega(\varrho)$ and put

$$w_T(t) = \begin{cases} w(t) & \text{if } 0 \leq t \leq T, \\ 0 & \text{if } t > T, \end{cases}$$

and

$$F_T w(t) = \begin{cases} (Fw)(t) & \text{if } 0 \leq t \leq T, \\ 0 & \text{if } t > T. \end{cases}$$

Since F is causal, one has $F_T w = F_T w_T$. Consequently

$$\begin{aligned} \|Fw\|_{C(0,T)} &= \|F_T w\|_{C(0,\infty)} = \|F_T w_T\|_{C(0,\infty)} \\ &\leq \|Fw_T\|_{C(0,\infty)} \leq q\|w_T\|_{C(-\eta,\infty)} = q\|w\|_{C(-\eta,T)}. \end{aligned}$$

Furthermore, take $v \in \Omega(\varrho)$ and put

$$v_T(t) = \begin{cases} v(t) & \text{if } 0 \leq t \leq T, \\ 0 & \text{if } t > T. \end{cases}$$

and

$$\delta = \|w_T - v_T\|_{C(-\eta,T)} \quad \text{and} \quad \epsilon = \|Fw_T - Fv_T\|_{C(-\eta,\infty)}.$$

We have $\|w_T - v_T\|_{C(-\eta,\infty)} = \delta$. Since F is continuous in $\Omega(\varrho)$ and

$$\|F_T w_T - F_T v_T\|_{C(-\eta,\infty)} \leq \|Fw_T - Fv_T\|_{C(-\eta,\infty)}$$

we prove the continuity of F in $\Omega(\varrho)$. This proves the result. □

A (mild) solution of problem (1.1), (1.2) is a continuous function $x(t)$ defined on $[-\eta,\infty)$, such that

$$x(t) = z(t) + \int_0^t G(t,t_1)([Fx](t_1) + f(t_1))dt_1 \quad (t \geq 0), \tag{1.4a}$$

$$x(t) = \phi(t) \in C(-\eta,0) \quad (-\eta \leq t \leq 0), \qquad 1.4b$$

where $G(t,t_1)$ is the fundamental solution of the linear equation

$$\dot{z}(t) = \int_0^\eta d_s R(t,s) z(t-s) \tag{1.5}$$

and $z(t)$ is a solution of the problem (1.5), (1.2). Again use the operator

$$\hat{G}f(t) = \int_0^t G(t,t_1)f(t_1)dt_1 \quad (f \in C(0,\infty)). \tag{1.6}$$

It is assumed that

$$\|z\|_{C(-\eta\,,\infty)} + \|\hat{G}\|_{C(0,\infty)}(q\varrho + \|f\|_{C(0,\infty)}) < \varrho \text{ if } \varrho < \infty, \tag{1.7a}$$

or

$$q\|\hat{G}\|_{C(0,\infty)} < 1, \text{ if } \varrho = \infty. \tag{1.7b}$$

Theorem 10.1.2. *Let F be a continuous causal mapping in $C(-\eta\,,\infty)$. Let conditions* (1.3) *and* (1.7) *hold. Then problem* (1.1), (1.2) *has a solution $x(t)$ satisfying the inequality*

$$\|x\|_{C(-\eta\,,\infty)} \le \frac{\|z\|_{C(-\eta\,,\infty)} + \|\hat{G}\|_{C(0,\infty)}\|f\|_{C(0,\infty)}}{1 - q\|\hat{G}\|_{C(0,\infty)}}. \tag{1.8}$$

Proof. Take a finite $T > 0$ and define on $\Omega_T(\varrho) = \Omega(\varrho) \cap C(-\eta\,,T)$ the mapping Φ by

$$\Phi w(t) = z(t) + \int_0^t G(t,s)([Fw](s) + f(s))ds \quad (0 \le t \le T; w \in \Omega_T(\varrho)),$$

and

$$\Phi w(t) = \phi(t) \text{ for } -\eta \le t \le 0.$$

Clearly, Φ maps $\Omega(\varrho)$ into $C(-\eta\,,T)$. Moreover, by (1.3) and Lemma 10.1.1, we obtain the inequality

$$\begin{aligned}&\|\Phi w\|_{C(-\eta\,,T)}\\ &\quad \le \max\{\|z\|_{C(0,T)} + \|\hat{G}\|_{C(0,\infty)}(q\|w\|_{C(-\eta\,,T)} + \|f\|_{C(0,\infty)}), \|\phi\|_{C(-\eta\,,0)}\}.\end{aligned}$$

But

$$\max\{\|z\|_{C(0,T)}, \|\phi\|_{C(-\eta\,,T)}\} = \|z\|_{C(-\eta\,,T)}.$$

So

$$\|\Phi w\|_{C(-\eta\,,T)} \le \|z\|_{C(-\eta\,,T)} + \|\hat{G}\|_{C(0,\infty)}(q\|w\|_{C(-\eta\,,T)} + \|f\|_{C(0,\infty)}).$$

According to (1.7) Φ maps $\Omega_T(\varrho)$ into itself. Taking into account that Φ is compact we prove the existence of solutions.

Furthermore, we have

$$\begin{aligned}\|x\|_{C(-\eta\,,T)} &= \|\Phi x\|_{C(-\eta\,,T)}\\ &\le \|z\|_{C(-\eta\,,T)} + \|\hat{G}\|_{C(0,T)}(q\|x\|_{C(-\eta\,,T)} + \|f\|_{C(0,\infty)}).\end{aligned}$$

Hence we easily obtain (1.8), completing the proof. □

Note that the Lipschitz condition

$$\|Fw - Fw_1\|_{C(0,\infty)} \le q\|w - w_1\|_{C(-\eta\,,\infty)} \;\; (w_1, w \in \Omega(\varrho)) \tag{1.9}$$

together with the Contraction Mapping theorem allows us easily to prove the existence and uniqueness of solutions. Namely, the following result is valid.

Theorem 10.1.3. *Let F be a continuous causal mapping in $C(-\eta\,,\infty)$. Let conditions* (1.7) *and* (1.9) *hold. Then problem* (1.1), (1.2) *has a unique solution $x \in \Omega(\varrho)$.*

Note that *in our considerations one can put $[Fx](t) = 0$ for $-\eta \le t < 0$.*

10.2 Stability of quasilinear equations

In the rest of this chapter the uniqueness of solutions is assumed.

Consider the equation

$$\dot{x}(t) = \int_0^{\eta} d_s R(t,s)x(t-s) + [Fx](t) \; (t \ge 0), \tag{2.1}$$

Recall that any causal mapping satisfies the condition $F0 \equiv 0$ (see Section 1.8).

Definition 10.2.1. Let F be a continuous causal mapping in $C(-\eta\,,\infty)$. Then the zero solution of (2.1) is said to be stable (in the Lyapunov sense), if for any $\epsilon > 0$, there exists a $\delta > 0$, such that the inequality $\|\phi\|_{C(-\eta,0)} \le \delta$ implies $\|x\|_{C(0,\infty)} \le \epsilon$ for any solution $x(t)$ of problem (2.1), (1.2).

The zero solution of (2.1) is said to be asymptotically stable, if it is stable, and there is an open set $\tilde{\Omega} \subseteq C(-\eta, 0)$, such that $\phi \in \tilde{\Omega}$ implies $x(t) \to 0$ as $t \to \infty$. Besides, $\tilde{\Omega}$ is called the region of attraction of the zero solution. If the zero solution of (2.1) is asymptotically stable and $\tilde{\Omega} = C(-\eta, 0)$, then it is globally asymptotically stable.

The zero solution of (2.1) is exponentially stable, if there are positive constants ν, m_0 and r_0, such that the condition $\|\phi\|_{C(-\eta\,,0)} \le r_0$ implies the relation

$$\|x(t)\|_n \le m_0\|\phi\|_{C(-\eta\,,0)}\, e^{-\nu t} \;\; (t \ge 0).$$

The zero solution of (2.1) is globally exponentially stable if it is exponentially stable and the region of attraction coincides with $C(-\eta\,,0)$. That is, $r_0 = \infty$.

Theorem 10.2.2. *Let the conditions* (1.3) *and*

$$q\|\hat{G}\|_{C(0,\infty)} < 1 \tag{2.2}$$

hold. Then the zero solution to (2.1) *is stable.*

This result immediately follows from Theorem 10.1.2.

Clearly,

$$\|\hat{G}\|_{C(0,\infty)} \le \sup_{t\ge 0} \int_0^t \|G(t,t_1)\|_n dt_1. \tag{2.3}$$

Thus the previous theorem implies

Corollary 10.2.3. *Let the conditions* (1.3) *and*

$$\sup_{t\ge 0} \int_0^t \|G(t,t_1)\| dt_1 < \frac{1}{q}$$

hold. Then the zero solution to (2.1) *is stable*

By Theorem 10.1.2, under conditions (1.3) and (2.2) we have the solution estimate

$$\|x\|_{C(-\eta,\infty)} \le \frac{\|z\|_{C(-\eta,\infty)}}{1 - q\|\hat{G}\|_{C(0,\infty)}} \tag{2.4}$$

provided

$$\|z\|_{C(-\eta,\infty)} < \varrho(1 - q\|\hat{G}\|_{C(0,\infty)}).$$

Since (1.5) is assumed to be stable, there is a constant c_0, such that

$$\|z\|_{C(-\eta,\infty)} \le c_0\|\phi\|_{C(-\eta,0)}. \tag{2.5}$$

Thus the inequality

$$c_0\|\phi\|_{C(-\eta,0)} \le \varrho(1 - q\|\hat{G}\|_{C(0,\infty)}) \tag{2.6}$$

gives us a bound for the region of attraction.

Furthermore, if the condition

$$\lim_{\|w\|_{C(-\eta,\infty)}\to 0} \frac{\|Fw\|_{C(0,\infty)}}{\|w\|_{C(-\eta,\infty)}} = 0 \tag{2.7}$$

holds, then equation (2.1) will be called *a quasilinear equation.*

Theorem 10.2.4 (Stability in the linear approximation). *Let* $\|\hat{G}\|_{C(0,\infty)} < \infty$ *and equation* (2.1) *be quasilinear. Then the zero solution to equation* (2.1) *is stable.*

Proof. From (2.7) it follows that for any $\varrho > 0$, there is a $q > 0$, such that (1.3) holds, and $q = q(\varrho) \to 0$ as $\varrho \to 0$. Take ϱ in such a way that the condition $q\|\hat{G}\|_{C(0,\infty)} < 1$ is fulfilled. Now the required result is due to the previous theorem. □

For instance, assume that

$$\|Fw(t)\|_n \le \sum_{k=1}^m \int_0^\eta \|w(t-s)\|_n^{p_k} d\mu_k(s) \quad (w \in C(-\eta,\infty)), \tag{2.8}$$

where $\mu_k(s)$ are nondecreasing functions, and $p_k = \text{const} > 1$. Then

$$\|Fw\|_{C(0,\infty)} \le \sum_{k=1}^{m} \operatorname{var}(\mu_k)\|w\|^{p_k}_{C(-\eta,\infty)}.$$

So (2.7) is valid. Moreover, for any $\varrho > 0$,

$$\|Fw(t)\|_n \le \sum_{k=1}^{m} \int_0^\eta \|w(t-s)\|_n d\mu_k(s)\varrho^{p_k-1} \quad (w \in \Omega(\varrho)).$$

So condition (1.3) holds with

$$q = q(\varrho) = \sum_{k=1}^{m} \varrho^{p_k-1}\operatorname{var}(\mu_k). \tag{2.9}$$

Furthermore, consider the following equation with an autonomous linear part:

$$\dot{x}(t) = \int_0^\eta dR_0(s)x(t-s) + [Fx](t), \tag{2.10}$$

where $R_0(\tau)$ is an $n \times n$-matrix-valued function defined on $[0,\eta\,]$ and having a bounded variation. Let $G_0(t)$ be the fundamental solution of the equation

$$\dot{z}(t) = \int_0^\eta dR_0(s)z(t-s). \tag{2.11}$$

Put

$$\hat{G}_0 f(t) = \int_0^t G_0(t-t_1)f(t_1)dt_1, \tag{2.12}$$

for a $f \in C(0,\infty)$. Theorem 10.2.2 implies that, if the conditions (1.3) and

$$q\|\hat{G}_0\|_{C(0,\infty)} < 1$$

hold, then the zero solution to equation (2.10) is stable.

Hence we arrive at the following result.

Corollary 10.2.5. *Let the conditions* (1.3) *and*

$$\|G_0\|_{L^1(0,\infty)} < \frac{1}{q} \tag{2.13}$$

hold. Then the zero solution to equation (2.10) *is stable.*

Recall that some estimates for $\|G_0\|_{C(0,\infty)}$ and $\|G_0\|_{L^1(0,\infty)}$ can be found in Sections 4.4 and 4.8 (in the general case) and Section 4.7 (in the case of systems with one delay).

10.3 Absolute L^p-stability

Let F be a continuous causal mapping in $L^p(-\eta\,,\infty)$ for some $p \geq 1$. That is, F maps $L^p(-\eta\,,\infty)$ into itself continuously in the norm of $L^p(-\eta\,,\infty)$, and for all $\tau > -\eta$ we have $P_\tau F = P_\tau F P_\tau$, where P_τ are the projections defined by

$$(P_\tau w)(t) = \begin{cases} w(t) & \text{if } -\eta \leq t \leq \tau, \\ 0 & \text{if } \tau < t < \infty, \end{cases} \qquad (w \in L^p(-\eta\,,\infty))$$

and $P_\infty = I$. Consider equation (2.10) assuming that the inequality

$$\|Fw\|_{L^p(0,\infty)} \leq q_p \|w\|_{L^p(-\eta\,,\infty)} \quad (w \in L^p(-\eta\,,\infty)) \tag{3.1}$$

is fulfilled with a constant q_p.

Repeating the proof of Lemma 10.1.1 we arrive at the following result.

Lemma 10.3.1. *Let F be a continuous causal mapping in $L^p(-\eta\,,\infty)$ for some $p \geq 1$. Let condition* (3.1) *hold. Then*

$$\|Fw\|_{L^p(0,T)} \leq q_p \|w\|_{L^p(-\eta\,,T)} \quad (w \in L^p(-\eta\,,T))$$

for all $T > 0$.

Let $\hat{G}_0$ be defined in space $L^p(0,\infty)$ by (2.12).

Theorem 10.3.2. *Let F be a continuous causal mapping in $L^p(-\eta\,,\infty)$ for some $p \geq 1$. Let the conditions* (3.1) *and*

$$q_p \|\hat{G}_0\|_{L^p(0,\infty)} < 1 \tag{3.2}$$

hold. Then problem (2.10), (1.2) *has a solution $x(t) \in L^p(-\eta\,,\infty)$ satisfying the inequality*

$$\|x\|_{L^p(-\eta\,,\infty)} \leq \frac{\|z\|_{L^p(-\eta\,,\infty)}}{1 - q_p \|\hat{G}_0\|_{L^p(0,\infty)}}, \tag{3.3}$$

where $z(t)$ is a solution of the linear problem (2.11), (1.2).

The proof of this theorem is similar to the proof of Theorem 10.1.2 with the replacement of $C(0,T)$ by $L^p(0,T)$.

The Lipschitz condition

$$\|Fw - Fw_1\|_{L^p(0,\infty)} \leq q_p \|w - w_1\|_{L^p(-\eta\,,\infty)} \quad (w_1, w \in L^p(0,\infty)) \tag{3.4}$$

together with the Contraction Mapping theorem also allows us easily to prove the existence and uniqueness of solutions. Specifically the following result is valid.

Theorem 10.3.3. *Let F be a continuous causal mapping in $L^p(-\eta\,,\infty)$ for some $p \geq 1$. Let conditions* (3.2) *and* (3.4) *hold. Then problem* (2.10), (1.2) *has a unique (continuous) solution $x \in L^p(-\eta\,,\infty)$.*

Definition 10.3.4. The zero solution to equation (2.10) is said to be absolutely L^p-stable in the class of the nonlinearities satisfying (3.1), if there is a positive constant m_0 independent of the specific form of functions F (but dependent on q_p), such that

$$\|x\|_{L^p(-\eta\,,\infty)} \leq m_0\|\phi\|_{C(-\eta\,,0)} \tag{3.5}$$

for any solution $x(t)$ of problem (2.10), (1.2).

From Theorem 10.3.2 it follows that the zero solution to equation (2.10) is absolutely L^p-stable in the class of nonlinearities satisfying (3.1), provided condition (3.2) holds.

According to the well-known property of convolutions (see Section 1.3) we have

$$\|\hat{G}_0\|_{L^p(0,\infty)} \leq \|G_0\|_{L^1(0,\infty)} \tag{3.6}$$

We thus arrive at the following result.

Corollary 10.3.5. *The zero solution to equation* (2.10) *is absolutely L^p-stable in the class of nonlinearities satisfying* (3.1) *provided* $q_p\|G_0\|_{L^1(0,\infty)} < 1$.

Recall that

$$K(z) = Iz - \int_0^{\eta} \exp(-zs)dR_0(s) \ (z \in \mathbb{C}).$$

It is assumed that all the characteristic values of K are in C_-. So the autonomous linear equation (2.11) is exponentially stable. According to Lemma 4.4.1 we have the inequality

$$\|\hat{G}_0\|_{L^2(0,\infty)} \leq \theta(K),$$

where

$$\theta(K) := \sup_{-2\ \mathrm{var}(R_0)\leq\omega\leq 2\ \mathrm{var}(R_0)} \|K^{-1}(i\omega)\|_n.$$

So due to Theorem 10.3.2 we get

Corollary 10.3.6. *The zero solution to equation* (2.10) *is absolutely L^2-stable in the class of nonlinearities satisfying*

$$\|Fw\|_{L^2(0,\infty)} \leq q_2\|w\|_{L^2(-\eta\,,\infty)} \ (w \in L^2(-\eta\,,\infty)), \tag{3.7}$$

provided

$$q_2\theta(K) < 1. \tag{3.8}$$

Moreover, any its solutions satisfies the inequality

$$\|x\|_{L^2(-\eta\,,\infty)} \leq \frac{\|z\|_{L^2(-\eta\,,\infty)}}{1 - q_2\theta(K)}, \tag{3.9}$$

where $z(t)$ is a solution of the linear problem (2.11), (1.2).

Now we can apply the bounds for $\theta(K)$ from Sections 4.3, 4.7 and 4.8.

The following lemma shows that the notion of the L^2-absolute stability of (2.10) is stronger than the notion of the asymptotic (absolute) stability.

Lemma 10.3.7. *If the zero solution to equation* (2.10) *is absolutely L^2-stable in the class of nonlinearities* (3.7), *then the zero solution to* (2.10) *is asymptotically stabile.*

Proof. Indeed, assume that a solution x of (2.10) is in $L^2(0,\infty)$ and note that from (2.10) and (3.7) it follows that $\|\dot{x}\|_{L^2(0,\infty)} \leq (\mathrm{var}(R_0) + q_2)\|x\|_{L^2(-\eta,\infty)}$. Thus

$$\|x(t)\|_n^2 = -\int_t^\infty \frac{d}{ds}\|x(s)\|_n^2 ds \leq 2\int_t^\infty \|x(s)\|_n \|\dot{x}(s)\|_n ds$$
$$\leq 2\left(\int_t^\infty \|x(s)\|_n^2 ds\right)^{1/2}\left(\int_t^\infty \|\dot{x}(s)\|_n^2 ds\right)^{1/2} \to 0 \quad \text{as } t \to \infty,$$

as claimed. □

10.4 Mappings defined on $\Omega(\varrho) \cap L^2$

In this section we investigate a causal mapping F acting in space $L^2(-\eta,\infty) = L^2([-\eta,\infty),\mathbb{C}^n)$ and satisfying the condition

$$\|Ff\|_{L^2(0,\infty)} \leq q_2\|f\|_{L^2(-\eta,\infty)} \quad (f \in \Omega(\varrho) \cap L^2(-\eta,\infty)) \tag{4.1}$$

with a positive constant q_2. Such a condition enables us to derive stability conditions, sharper than (2.2).

For example, let there be a nondecreasing function $\nu(s) = \nu(\varrho, s)$, defined on $[0,\eta]$, such that

$$\|[Ff](t)\|_n \leq \int_0^\eta \|f(t-s)\|_n d\nu(s) \quad (t \geq 0;\ f \in \Omega(\varrho)). \tag{4.2}$$

Let us show that condition (4.2) implies the inequality

$$\|Ff\|_{L^2(0,\infty)} \leq \mathrm{var}(\nu)\ \|f\|_{L^2(-\eta,\infty)} \quad (f \in \Omega(\varrho) \cap L^2(-\eta,\infty)). \tag{4.3}$$

Indeed, introduce in the space of scalar functions $L^2([-\eta,\infty);\mathbb{C})$ the operator $\hat{E}_\nu$ by

$$(\hat{E}_\nu w)(t) = \int_0^\eta w(t-\tau)d\nu(\tau) \quad (w \in L^2([-\eta,\infty);\mathbb{C}).$$

Then

$$\|\hat{E}_\nu w\|^2_{L^2(0,\infty)} = \int_0^\infty \left| \int_0^\eta w(t-\tau)d\nu(\tau) \right|^2 dt$$
$$= \int_0^\infty \left| \int_0^\eta w(t-\tau)d\nu(\tau) \int_0^\eta w(t-\tau_1)d\nu(\tau_1) \right| dt$$
$$\le \int_0^\infty \int_0^\eta |w(t-\tau)|d\nu(\tau) \int_0^\eta |w(t-\tau_1)|d\nu(\tau_1)|dt.$$

Hence

$$\|\hat{E}_\nu w\|^2_{L^2(0,\infty)} \le \int_0^\eta \int_0^\eta \int_0^\infty |w(t-\tau)||w(t-\tau_1)|dt \ \ d\nu(\tau_1)\, d\nu(\tau).$$

But by the Schwarz inequality

$$\int_0^\infty |w(t-\tau)||w(t-\tau_1)|dt \le \sqrt{\int_0^\infty |w(t-\tau)|^2 dt} \sqrt{\int_0^\infty |w(t-\tau_1)|^2 dt}$$
$$\le \int_{-\eta}^\infty |w(t)|^2 dt.$$

Now simple calculations show that

$$\|\hat{E}_\nu w\|_{L^2(0,\infty)} \le \mathrm{var}(\nu)\|w\|_{L^2(-\eta\,,\infty)}.$$

Hence (4.2) with $w(t) = \|f(t)\|_n$ implies (4.3).

Theorem 10.4.1. *Let F be a continuous causal mapping in $L^2(-\eta\,,\infty)$. Let conditions* (4.1) *and* (3.8) *hold. Then the zero solution to* (2.10) *is L^2-stable. Namely, there are constants $c_0 \ge 0$ and $r_0 \in (0, \varrho]$, such that*

$$\|x\|_{L^2(-\eta\,,\infty)} \le c_0\|\phi\|_{C(-\eta\,,0)} \tag{4.4}$$

for a solution $x(t)$ of problem (2.10), (1.2) *provided*

$$\|\phi\|_{C(-\eta\,,0)} < r_0. \tag{4.5}$$

Proof. First let $\varrho = \infty$. Since the linear equation (2.11) is L^2-stable, we can write

$$\|z\|_{L^2(-\eta\,,\infty)} \le c_1\|\phi\|_{C(-\eta\,,0)} \ \ (c_1 = \mathrm{const}). \tag{4.6}$$

By (3.9) we thus have (4.4). So in the case $\varrho = \infty$, the theorem is proved.

Furthermore, recall that

$$(E_0 f)(t) = \int_0^\eta dR_0(\tau) f(t-\tau).$$

By Lemma 1.12.1

$$\|E_0 f\|_{L^2(0,\infty)} \le \mathrm{var}(R_0)\|f\|_{L^2(-\eta\,,\infty)}.$$

Now from (2.10) and (4.1) in the case $\varrho = \infty$ it follows that

$$\|\dot{x}\|_{L^2(0,\infty)} \le (\mathrm{var}(R_0) + q_2)\|x\|_{L^2(-\eta\,,\infty)}.$$

Or according to (4.4),

$$\|\dot{x}\|_{L^2(0,\infty)} \le (\mathrm{var}(R_0) + q_2)c_0\|\phi\|_{C(-\eta\,,0)}. \tag{4.7}$$

Recall that by Lemma 4.4.6, if $f \in L^2(0,\infty)$ and $\dot{f} \in L^2(0,\infty)$, then

$$\|f\|^2_{C(0,\infty)} \le 2\|f\|_{L^2(0,\infty)}\|\dot{f}\|_{L^2(0,\infty)}.$$

This and (4.7) imply the inequality

$$\|x\|_{C(0,\infty)} \le c_2\|\phi\|_{C(-\eta\,,0)} \ \ (c_2 = c_0\sqrt{2}). \tag{4.8}$$

Now let $\varrho < \infty$. By the Urysohn theorem, (see Section 1.1), there is a continuous scalar-valued function ψ_ϱ defined on $C(0,\infty)$, such that

$$\psi_\varrho(f) = \begin{cases} 1 & \text{if } \|f\|_{C(0,\infty)} < \varrho, \\ 0 & \text{if } \|f\|_{C(0,\infty)} \ge \varrho. \end{cases}$$

Put $F_\varrho f = \psi_\varrho(f)Ff$. Clearly, F_ϱ satisfies (4.1) for all $f \in L^2(-\eta\,,\infty)$. Consider the equation

$$\dot{x} = E_0 x + F_\varrho x. \tag{4.9}$$

Denote the solution of problems (4.9), (1.2) by x_ϱ. According to (4.4) and (4.8) we have

$$\|x_\varrho\|_{L^2(-\eta\,,\infty)} \le c_0\|\phi\|_{C(-\eta\,,0)} \quad \text{and} \quad \|x_\varrho\|_{C(0,\infty)} \le c_2\|\phi\|_{C(-\eta\,,0)}.$$

If we take

$$\|\phi\|_{C(-\eta\,,0)} < r_0 = \frac{\varrho}{c_2},$$

then $x_\varrho \in \Omega(\varrho)$. So the solutions of equations (2.10) and (4.8) coincide. This proves the theorem. □

10.5 Exponential stability

Let $\tilde{F}$ be a continuous causal operator in $C(-\eta\,,\infty)$. Consider the equation

$$\dot{x} = \tilde{F}x \tag{5.1}$$

and substitute

$$x(t) = y_\epsilon(t)e^{-\epsilon t} \tag{5.2}$$

with an $\epsilon > 0$ into (5.1). Then we obtain the equation

$$\dot{y}_\epsilon = \epsilon y_\epsilon + e^{\epsilon t}\tilde{F}(e^{-\epsilon t}y_\epsilon). \tag{5.3}$$

Lemma 10.5.1. *For an $\epsilon > 0$, let the zero solution of equation* (5.3) *be stable in the Lyaponov sense. Then the zero solution of equation* (5.1) *is exponential stable.*

Proof. If $\|\phi\|_{C(-\eta,0)}$ is sufficiently small, we have

$$\|y_\epsilon(t)\|_n \le m_0\|\phi\|_{C(-\eta,0)} \quad (t \ge 0)$$

for a solution of (5.3). Now (5.2) implies the result. □

Recall that $\Omega(\varrho) := \{v \in C(-\eta,\infty) : \|v\|_{C(-\eta,\infty)} \le \varrho\}$ for a positive $\varrho \le \infty$.

Let a continuous causal mapping F in $C(-\eta,\infty)$ satisfy condition (1.3), *for an $\varrho \le \infty$. Then we will say that F has the ϵ-property, if for any $f \in \Omega(\varrho)$ we have*

$$\lim_{\epsilon\to 0} \|e^{\epsilon t}F(e^{-\epsilon t}f)\|_{C(0,\infty)} \le q\|f\|_{C(0,\infty)}. \tag{5.4}$$

Here

$$\|e^{\epsilon t}F(e^{-\epsilon t}f)\|_{C(0,\infty)} = \sup_{t\ge 0} \|e^{\epsilon t}[F(e^{-\epsilon t}f)](t)\|_n.$$

For example, if condition (4.2) is fulfilled, then

$$\begin{aligned}\|e^{\epsilon t}[F(e^{-\epsilon t}f)](t)\|_n &\le e^{t\epsilon}\int_0^\eta e^{-\epsilon(t-s)}\|f(t-s)\|_n d\nu \\ &\le e^{\epsilon\eta}\int_0^\eta \|f(t-s)\|_n d\nu \quad (f \in \Omega(\varrho)),\end{aligned}$$

and thus condition (5.4) holds.

Theorem 10.5.2. *Let conditions* (1.3), (2.2) *and* (5.4) *hold. Then the zero solution to* (2.1) *is exponentially stable.*

Proof. Substituting (5.2) with a sufficiently small $\epsilon > 0$ into (2.1), we obtain the equation

$$\dot{y}_\epsilon - \epsilon y_\epsilon = E_{\epsilon,R}y_\epsilon + F_\epsilon y_\epsilon, \tag{5.5}$$

where

$$(E_{\epsilon,R}f)(t) = \int_0^\eta e^{\epsilon\tau}d_\tau R(t,\tau)f(t-\tau) \quad \text{and} \quad [F_\epsilon f](t) = e^{\epsilon t}[F(e^{-\epsilon t}f)](t).$$

By (5.4) we have

$$\|F_\epsilon f\|_{C(0,\infty)} \le a(\epsilon)\|Ff\|_{C(0,\infty)} \tag{5.6}$$

where $a(\epsilon) \to l \le 1$ as $\epsilon \to 0$. Let G_ϵ be the fundamental solution of the linear equation

$$\dot{y} - \epsilon y = E_{\epsilon,R} y. \tag{5.7}$$

Put

$$\hat{G}_\epsilon f(t) = \int_0^t G_\epsilon(t, t_1) f(t_1) dt_1 \quad (f \in C(0,\infty)).$$

It is simple to see that $\hat{G}_\epsilon \to \hat{G}$ as $\epsilon \to 0$. Taking ϵ sufficiently small and applying Theorem 10.2.2 according to (5.6) we can assert that equation (5.5) is stable in the Lyapunov sense. Now Lemma 10.5.1 proves the exponential stability. □

10.6 Nonlinear equations "close" to ordinary differential ones

Consider the equation

$$\dot{y}(t) = A(t)y(t) + [Fy](t) \quad (t \ge 0), \tag{6.1}$$

where $A(t)$ is a piece-wise continuous matrix-valued function and F is a continuous causal mapping F in $C(-\eta\,,\infty)$. Recall the $\|.\|_n$ means the Euclidean norm for vectors and the corresponding operator (spectral) norm for matrices.

Lemma 10.6.1. *Let F be a continuous causal mapping in $C(-\eta\,,\infty)$ satisfying condition* (1.3), *and the evolution operator $U(t,s)$ $(t \ge s \ge 0)$ of the equation*

$$\dot{y} = A(t)y \ (t > 0) \tag{6.2}$$

satisfy the condition

$$\nu_\infty := \sup_{t \ge 0} \int_0^t \|U(t,s)\|_n ds < \frac{1}{q}. \tag{6.3}$$

Then the zero solution of equation (6.1) *is stable. Moreover, a solution y of problem* (6.1), (1.2) *satisfies the inequality*

$$\|y\|_{C(0,\infty)} \le \frac{\sup_{t\ge 0} \|U(t,0)\phi(0)\|_n + q\nu_\infty \|\phi\|_{C(-\eta\,,0)}}{1 - q\nu_\infty}$$

provided

$$\frac{\sup_{t\ge 0} \|U(t,0)\phi(0)\|_n + q\nu_\infty \|\phi\|_{C(-\eta\,,0)}}{1 - q\nu_\infty} < \varrho. \tag{6.4}$$

Proof. Rewrite (6.1) as

$$x(t) = U(t,0)\phi(0) + \int_0^t U(t,s)(Fx)(s)ds.$$

So

$$\|x(t)\|_n \le \|U(t,0)\phi(0)\|_n + \int_0^t \|U(t,s)\|_n \|Fx(s)\|_n ds.$$

According to (6.4) and continuity of solutions, there is a $T > 0$, such that

$$\|x(t)\|_n < \varrho \ \ (t \le T).$$

Since F is causal, by (1.3)

$$\|Fx\|_{C(0,T)} \le q\|x\|_{C(-\eta,T)}.$$

Thus

$$\|x\|_{C(0,T)} \le \sup_{0\le t\le T} \|U(t,0)\phi(0)\|_n + q\|x\|_{C(-\eta,T)} \sup_{0\le t\le T} \int_0^t \|U(t,s)\|_n ds.$$

Consequently

$$\begin{aligned}\|x\|_{C(0,T)} &\le \sup_{t\ge 0} \|U(t,0)\phi(0)\|_n + q\|x\|_{C(-\eta,T)}\nu_\infty \\ &\le \sup_{t\ge 0} \|U(t,0)\phi(0)\|_n + q\nu_\infty(\|x\|_{C(0,T)} + \|\phi\|_{C(-\eta,0)}).\end{aligned}$$

Hence,

$$\|x\|_{C(0,T)} \le \frac{\sup_{t\ge 0} \|U(t,0)\phi(0)\|_n + q\nu_\infty \|\phi\|_{C(-\eta,0)}}{1 - q\nu_\infty}.$$

Now condition (6.4) enables us to extend this inequality to the whole half-line, as claimed. □

Due to the latter theorem we arrive at our next result.

Corollary 10.6.2. *Under the hypothesis of Theorem* 10.6.1, *let F have the ϵ-property* (5.4). *Then the zero solution of equation* (6.1) *is exponentially stable.*

Note that, if

$$((A(t) + A^*(t))h, h)_{C^n} \le -2\alpha(t)(h,h)_{C^n} \ \ (h \in \mathbb{C}^n, t \ge 0)$$

with a positive a piece-wise continuous function $\alpha(t)$, then

$$\|U(t,s)\|_n \le e^{-\int_s^t \alpha(t_1)dt_1}$$

So if α has the property

$$\hat{\nu}_\infty := \sup_{t\ge 0} \int_0^t e^{-\int_s^t \alpha(t_1)dt_1} ds < \infty,$$

then $\nu_\infty \leq \hat{\nu}_\infty$. For instance, under the condition

$$\alpha_0 := \inf_t \alpha(t) > 0,$$

we deduce that

$$\hat{\nu}_\infty \leq \sup_{t\geq 0} \int_0^t e^{-\alpha_0(t-s)} ds = 1/\alpha_0.$$

In particular, if $A(t) \equiv A_0$ is a constant matrix, then

$$U(t,s) = e^{(t-s)A_0} \quad (t \geq s \geq 0).$$

As it is shown in Section 2.5,

$$\|e^{A_0 t}\|_n \leq e^{\alpha(A_0)t} \sum_{k=0}^{n-1} \frac{t^k g^k(A_0)}{(k!)^{3/2}} \quad (t \geq 0),$$

where $\alpha(A_0) = \max_k \operatorname{Re} \lambda_k(A_0)$ and therefore,

$$\nu_\infty = \|e^{A_0 t}\|_{L^1(0,\infty)} \leq \nu_{A_0}, \quad \text{where} \quad \nu_{A_0} := \sum_{k=0}^{n-1} \frac{g^k(A_0)}{\sqrt{k!}|\alpha(A_0)|^{k+1}}.$$

Now Lemma 10.6.1 and Corollary 10.6.2 imply the following result.

Corollary 10.6.3. *Let A_0 be a Hurwitzian matrix and the conditions* (1.3) *and*

$$q\nu_{A_0} < 1$$

hold. Then the zero solution of the equation

$$\dot{x}(t) = A_0 x(t) + (Fx)(t) \tag{6.5}$$

is stable. If, in addition, F has the ϵ-property (5.4)*, then the zero solution of equation* (6.5) *is exponentially stable.*

10.7 Applications of the generalized norm

Again consider equation (2.1). Let $r_{jk}(t,s)$ be the entries of $R(t,s)$. Rewrite (2.1) in the form of the coupled system

$$\dot{x}_j(t) + \sum_{k=1}^{n} \int_0^\eta x_j(t-s) d_s r_{jk}(t,s) = [Fx]_j(t) \quad (t \geq 0;\ j = 1, \ldots, n), \tag{7.1}$$

where $(x(t) = (x_k(t))_{k=1}^n$, $[Fw]_j(t)$ mean the coordinates of the vector function $Fw(t)$ with a $w \in C([-\eta, \infty), \mathbb{C}^n)$.

The inequalities for real vectors or vector functions are understood below in the coordinate-wise sense.

Furthermore, let $\hat{\rho} := (\rho_1, \dots, \rho_n)$ be a vector with positive coordinates $\rho_j < \infty$. We need the following set:

$$\tilde{\Omega}(\hat{\rho}) := \{v(t) = (v_j(t)) \in C([-\eta\,,\infty), \mathbb{C}^n) : \|v_j\|_{C([-\eta\,,\infty),\mathbb{C})} \le \rho_j;\ \ j = 1, \dots, n\}.$$

If we introduce in $C([a,b], \mathbb{C}^n)$ the generalized norm as the vector

$$M_{[a,b]}(v) := (\|v_j\|_{C([a,b],\mathbb{C})})_{j=1}^n \ \ (v(t) = (v_j(t)) \in C([a,b], \mathbb{C}^n))$$

(see Section 1.7), then we can write down

$$\tilde{\Omega}(\hat{\rho}) := \{v \in C([-\eta\,,\infty), \mathbb{C}^n) : M_{[-\eta\,,\infty)}(v) \le \hat{\rho}\}.$$

It is assumed that F satisfies the following condition: there are non-negative constants ν_{jk} $(j,k = 1, \dots, n)$, such that for any

$$w(t) = (w_j(t))_{j=1}^n \in \tilde{\Omega}(\hat{\rho}),$$

the inequalities

$$\|[Fw]_j\|_{C([0,\infty),\mathbb{C})} \le \sum_{k=1}^n \nu_{jk}\|w_k\|_{C([-\eta\,,\infty),\mathbb{C})} \ \ (j = 1, \dots, n) \tag{7.2}$$

hold. In other words,

$$M_{[0,\infty)}(Fw) \le \Lambda(F) M_{[-\eta\,,\infty)}(w) \ \ (w \in \tilde{\Omega}(\hat{\rho})), \tag{7.3}$$

where $\Lambda(F)$ is the matrix defined by

$$\Lambda(F) = (\nu_{jk})_{j,k}^n. \tag{7.4}$$

Lemma 10.7.1. *Let F be a continuous causal mapping in $C(-\eta\,,\infty)$ satisfying condition* (7.3). *Then*

$$M_{[0,T]}(Fw) \le \Lambda(F) M_{[-\eta\,,T]}(w) \ \ (w \in \tilde{\Omega}(\hat{\rho}) \cap C([-\eta\,,T]), \mathbb{C}^n))$$

for all $T > 0$.

Proof. Again put

$$w_T(t) = \begin{cases} w(t) & \text{if } 0 \le t \le T, \\ 0 & \text{if } t > T, \end{cases}$$

and

$$F_T w(t) = \begin{cases} (Fw)(t) & \text{if } 0 \le t \le T, \\ 0 & \text{if } t > T. \end{cases}$$

Take into account that $F_T w = F_T w_T$. Then

$$\begin{aligned} M_{[0,T]}(Fw) &= M_{[0,\infty)}(F_T w) = M_{[0,\infty)}(F_T w_T) \\ &\leq M_{[0,\infty)}(F w_T) \leq \Lambda(F) M_{[-\eta\,,\infty)}(w_T) = \Lambda(F) M_{[-\eta\,,T]}(w). \end{aligned}$$

This proves the required result. □

It is also assumed that the entries $G_{jk}(t,s)$ of the fundamental solution $G(t,s)$ of equation (1.5) satisfy the conditions

$$\gamma_{jk} := \sup_{t\geq 0} \int_0^\infty |G_{jk}(t,s)|ds < \infty. \tag{7.5}$$

Denote by $\hat{\gamma}$ the matrix with the entries γ_{jk}:

$$\hat{\gamma} = (\gamma_{jk})_{j,k}^n.$$

Theorem 10.7.2. *Let the condition* (7.2) *and* (7.5) *hold. If, in addition, the spectral radius of the matrix* $Q = \hat{\gamma}\Lambda(F)$ *is less than one, then the zero solution of equation* (2.1) *is stable. Moreover, if a solution* z *of the linear problem* (1.5), (1.2) *satisfies the condition*

$$M_{[-\eta\,,\infty)}(z) + Q\hat{\rho} \leq \hat{\rho}\,, \tag{7.6}$$

then the solution $x(t)$ *of problem* (2.1), (1.2) *satisfies the inequality*

$$M_{[-\eta\,,\infty)}(x) \leq (I-Q)^{-1} M_{[-\eta\,,\infty)}(z). \tag{7.7}$$

Proof. Take a finite $T > 0$ and define on $\Omega_T(\hat{\rho}) = \tilde{\Omega}(\hat{\rho}) \cap C(-\eta\,,T)$ the mapping Φ by

$$\Phi w(t) = z(t) + \int_0^t G(t,t_1)[Fw](t_1)dt_1 \quad (0 \leq t \leq T; w \in \Omega_T(\hat{\rho})),$$

and

$$\Phi w(t) = \phi(t) \text{ for } -\eta \leq t \leq 0.$$

Then by (7.3) and Lemma 10.7.1,

$$M_{[-\eta\,,T]}(\Phi w) \leq M_{[-\eta\,,T]}(z) + \hat{\gamma}\Lambda(F) M_{[-\eta\,,T]}(w).$$

According to (7.6) Φ maps $\Omega_T(\hat{\rho})$ into itself. Taking into account that Φ is compact we prove the existence of solutions. Furthermore,

$$M_{[-\eta\,,T]}(x) = M_{[-\eta\,,T]}(\Phi x) \leq M_{[-\eta\,,T]}(z) + Q M_{[-\eta\,,T]}(x).$$

So

$$M_{[-\eta\,,T]}(x) \leq (I-Q)^{-1} M_{[-\eta\,,T]}(z).$$

Hence letting $T \to \infty$, we obtain (7.7), completing the proof. □

The Lipschitz condition

$$M_{[0,\infty)}(Fw - Fw_1) \le \Lambda(F) M_{[-\eta,\infty)}(w - w_1) \quad (w_1, w \in \tilde{\Omega}(\hat{\rho})) \tag{7.8}$$

together with the Generalized Contraction Mapping theorem (see Section 1.7) also allows us to prove the existence and uniqueness of solutions. Namely, the following result is valid.

Theorem 10.7.3. *Let conditions* (7.5) *and* (7.8) *hold. If, in addition, the spectral radius of the matrix* $Q = \hat{\gamma}\Lambda(F)$ *is less than one, then problem* (2.1), (1.2) *has a unique solution* $\tilde{x} \in \tilde{\Omega}(\hat{\rho})$*, provided* z *satisfies condition* (7.6). *Moreover, the zero solution of equation* (2.1) *is stable.*

The proof is left to the reader.

Note that one can use the well-known inequality

$$r_s(Q) \le \max_j \sum_{k=1}^{n} q_{jk}, \tag{7.9}$$

where q_{jk} are the entries of Q. About this inequality, as well as about other estimates for the matrix spectral radius see Section 2.4.

Let $\mu_j(s)$ be defined and nondecreasing on $[0, \eta\,]$. Now we are going to investigate the stability of the following nonlinear system with the diagonal linear part:

$$\dot{x}_j(t) + \int_0^{\eta} x_j(t-s)d\mu_j(s) = [F_j x](t) \quad (x(t) = (x_k(t))_{k=1}^n;\ j = 1, \dots, n;\ t \ge 0). \tag{7.10}$$

In this case $G_{jk}(t,s) = G_{jk}(t-s)$ and

$$G_{jk}(t) = 0 \ (j \ne k;\ j,k = 1, \dots, n;\ t \ge 0).$$

Suppose that,

$$\eta \ \operatorname{var}(\mu_j) < \frac{1}{e} \ (j = 1, \dots, n), \tag{7.11}$$

then $G_{jj}(s) \ge 0$ (see Section 11.3). Moreover, due to Lemma 4.6.3, we obtain the relations

$$\gamma_{jj} = \int_0^{\infty} G_{jj}(s)ds = \frac{1}{\operatorname{var}(\mu_j)}.$$

Thus

$$Q = (q_{jk})_{j,k=1}^n \text{ with the entries } q_{jk} = \frac{\nu_{jk}}{\operatorname{var}(\mu_j)}.$$

Now Theorem 10.7.2 and (7.9) imply our next result.

Corollary 10.7.4. *If the conditions* (7.2), (7.11) *and*

$$\frac{1}{\mathrm{var}(\mu_j)} \sum_{k=1}^{n} \nu_{jk} < 1 \ \ (j = 1, \ldots, n)$$

hold, then the zero solution of equation (7.10) *is stable.*

Furthermore, let us apply the generalized norm to equation (6.1). To this end we use the representation of the evolution operator by the exponential multiplicative integral:

$$U(t, s) = \int_{[s,t]}^{\leftarrow} e^{A(t_1)dt_1}$$

(see [14, Section III.1.5]), where the symbol

$$\int_{[s,t]}^{\leftarrow} e^{A(t_1)dt_1}$$

means the limit of the products

$$e^{A(\tau_n)\delta} e^{A(\tau_{n-1})\delta} \cdots e^{A(\tau_1\delta} e^{A(\tau_0)\delta}, \quad \left(\tau_k = \tau_k^{(n)} = s + \frac{(t-s)k}{n}, \delta = \frac{t-s}{n}\right)$$

as $n \to \infty$

Assume that $A(t) = (a_{jk}(t))$ is real and its norm is bounded on the positive half-line; so there are positive constants b_{jk} $(j \neq k)$ and real constants b_{jj}, such that

$$a_{jj}(t) \leq b_{jj} \quad \text{and} \quad |a_{jk}(t)| \leq b_{jk} \ \ (t \geq 0;\ j \neq k;\ j, k = 1, \ldots, n). \tag{7.12}$$

These inequalities imply

$$A(t) \leq B \quad \text{for all} \quad t \geq 0, \quad \text{where} \quad B = (b_{jk}).$$

For a matrix (c_{jk}) let $|C|$ mean the matrix $(|c_{jk}|)$. Then due to (7.12) we have

$$|e^{A(t)\delta}| \leq e^{B\delta}.$$

Hence

$$|U(t, s)| = \left| \int_{[s,t]}^{\leftarrow} e^{A(t_1)dt_1} \right| \leq \exp[(t-s)B] \ \ (t \geq s).$$

In the considered case $G(t, s) = U(t, s)$ and the entries $G_{jk}(t, s)$ of $G(t, s)$ satisfy

$$|G_{jk}(t, s)| \leq u_{jk}(t, s)$$

where $u_{jk}(t, s)$ are the entries of the matrix $\exp[(t-s)B]$.

Now we can apply Theorem 10.7.2 and estimates for $u_{jk}(t, s)$ from Section 2.6 to establish the stability condition for equation (6.1).

10.8 Systems with positive fundamental solutions

In this section all the considered functions and matrices are assumed to be real.

Let R^n_+ be the cone of vectors from $\mathbb{C}^n$ with non-negative coordinates. Denote by

$$C_+(a,b) = C([a,b], R^n_+)$$

the cone of vector functions from $C(a,b) = C([a,b], \mathbb{R}^n)$ with non-negative coordinates. The inequality $v \geq 0$ for a vector v means that $v \in R^n_+$. The inequality $f \geq 0$ for a function f means that $f \in C_+(a,b)$. The inequality $f_1 \geq f_2$ for functions f_1, f_2 means that $f_1 - f_2 \in C_+(a,b)$.

Consider the problem

$$\dot{x}(t) = \int_0^\eta d_s R(t,s) x(t-s) + [Fx](t), \tag{8.1}$$

$$x(t) = \phi(t) \in C_+(-\eta\,, 0) \quad (-\eta\ \leq t \leq 0), \tag{8.2}$$

where $R(t,s) = (r_{jk}(t,\tau))^n_{j,k=1}$ again is an $n \times n$-matrix-valued function defined on $[0,\infty) \times [0,\eta\,]$, which is piece-wise continuous in t for each τ, whose entries satisfy the condition

$$v_{jk} = \sup_{t \geq 0} = \text{var}\, r_{jk}(t,.) < \infty \quad (j,k = 1,\dots,n),$$

and F is a continuous causal mapping in $C(-\eta\,,\infty)$.

A solution of problem (8.1), (8.2) is understood as in Section 10.1. In this section the existence of solutions of problem (8.1), (8.2) is assumed.

Let the fundamental solution $G(t,s)$ of the equation

$$\dot{z}(t) = \int_0^\eta d_s R(t,s) z(t-s) \tag{8.3}$$

be a matrix function non-negative for all $t \geq s \geq 0$. Denote by K_η the subset of $C_+(-\eta\,,0)$, such that

$$\phi \in K_\eta \quad \text{implies} \quad z(t) \geq 0 \ (t \geq 0) \tag{8.4}$$

for a solution $z(t)$ of the linear problem (8.3), (8.2).

Recall that $F0 \equiv 0$.

Definition 10.8.1. Let F be a continuous causal mapping in $C(-\eta\,,\infty)$. Then the zero solution of (8.1) is said to be stable (in the Lyapunov sense) with respect to K_η, if for any $\epsilon > 0$, there exists a number $\delta > 0$, such that the conditions $\phi \in K_\eta$ and $\|\phi\|_{C(-\eta,0)} \leq \delta$ imply $\|x\|_{C(0,\infty)} \leq \epsilon$ for a solution $x(t)$ of problem (8.1), (8.2).

The zero solution of (8.1) is said to be asymptotically stable, with respect to K_η if it is stable with respect to K_η , and there is an open set $\omega_+ \subseteq K_\eta$, such that $\phi \in \omega_+$ implies $x(t) \to 0$ as $t \to \infty$.

The zero solution of (8.1) is exponentially stable with respect to K_η, if there are positive constants ν, m_0 and r_0, such that the conditions

$$\phi \in K_\eta \quad \text{and} \quad \|\phi\|_{C(-\eta\,,0)} \le r_0$$

imply the relation $\|x(t)\|_n \le m_0 \|\phi\|_{C(-\eta\,,0)}\, e^{-\nu t}$ $(t \ge 0)$ for any positive solution x. If $r_0 = \infty$, then the zero solution of (8.1) is globally exponentially stable with respect to K_η.

It is assumed that F satisfies the following conditions: there are linear causal non-negative bounded operators A_- and A_+, such that

$$A_- v \le Fv \le A_+ v \quad \text{for any} \quad v \in C_+(-\eta\,,\infty). \tag{8.5}$$

It particular, A_- can be the zero operator: $A_- v = 0$ for any positive v.

Theorem 10.8.2. *Let the fundamental solution $G(t,s)$ of equation (8.3) be a matrix function non-negative for all $t \ge s \ge 0$ and condition (8.5) hold. Then a solution $x(t)$ of problem (8.1), (8.2) with $\phi \in K_\eta$ satisfies the inequalities*

$$x_-(t) \le x(t) \le x_+(t) \ \ (t \ge 0), \tag{8.6}$$

where $x_+(t)$ is the (non-negative continuous) solution of the equation

$$\dot{x}(t) = \int_0^\eta d_s R(t,s) x(t-s) + [A_+ x](t) \tag{8.7}$$

and $x_-(t)$ is the solution of the equation

$$\dot{x}(t) = \int_0^\eta d_s R(t,s) x(t-s) + [A_- x](t) \tag{8.8}$$

with the same initial function $\phi \in K_\eta$.

Proof. Since $\phi(t) \ge 0$; we have $[Fx](s) \ge 0, s \le 0$. Take into account that

$$x(t) = z(t) + \int_0^t G(t,s)[Fx](s)ds.$$

So, there is a $T > 0$, such that

$$x(t) \ge 0 \ (0 \le t \le T) \text{ and therefore } x(t) \ge z(t) \ (0 \le t \le T).$$

But $z(t) \ge 0$ for all $t \ge 0$. Consequently, one can extend these inequalities to the whole positive half-line.

Now conditions (8.5) imply

$$x(t) \le z(t) + \int_0^t G(t,s)[A_+ x](s)ds \quad \text{and} \quad x(t) \ge z(t) + \int_0^t G(t,s)[A_- x](s)ds.$$

Hence due the abstract Gronwall Lemma (see Section 1.5), omitting simple calculations we have

$$x(t) \le x_+(t) \quad \text{and} \quad x(t) \ge x_-(t) \ (t \ge 0),$$

where $x_+(t)$ is a solution of the equation

$$x_+(t) = z(t) + \int_0^t G(t,s)[A_+ x_+](s)ds$$

and $x_-(t)$ is a solution of the equation

$$x_-(t) = z(t) + \int_0^t G(t,s)[A_- x_-]ds \ (t \ge 0).$$

According to the Variation of Constant formula these equations are equivalent to (8.7) and (8.8). □

Corollary 10.8.3. *Under the hypothesis of Theorem* 10.8.2, *let equation* (8.7) *be (asymptotically, exponentially) stable with respect to* K_η . *Then the zero solution to* (8.1) *is globally (asymptotically, exponentially) stable with respect to* K_η .

Conversely, let the zero solution to (8.1) *be globally (asymptotically, exponentially) stable with respect to* K_η . *Then equation* (8.8) *is (asymptotically, exponentially) stable with respect to* K_η .

Again use the generalized norm as the vector

$$M_{[a,b]}(v) := (\|v_j\|_{C([a,b],\mathbb{R})})_{j=1}^n \quad (v(t) = (v_j(t)) \in C([a,b],\mathbb{R}^n)).$$

Furthermore, let $\hat{\rho} := (\rho_1, \ldots, \rho_n)$ be a vector with positive coordinates $\rho_j < \infty$. Put

$$\Omega_+(\hat{\rho}) := \{v(t) = (v_j(t)) \in C([-\eta,\infty), R^n_+) : 0 \le v_j(t) \le \rho_j ; t \ge -\eta ; j = 1,\ldots,n\}.$$

So

$$\Omega_+(\hat{\rho}) := \{v \in C([-\eta\,,\infty), R_n^+) : M_{[-\eta\,,\infty)}(v) \le \hat{\rho}\}.$$

It should be noted that the Urysohn theorem and just proved theorem enable us, instead of conditions (8.3), to impose the following ones:

$$A_- v \le Fv \le A_+ v \text{ for any } v \in \Omega_+(\hat{\rho}).$$

10.9 The Nicholson-type system

In this section we explore the vector equation, which models cancer cell populations and other biological processes, cf. [8] and references given therein. Namely, we consider the system

$$\begin{aligned} \frac{dx_1}{dt} &= r(t)\left[-a_1 x_1(t) + b_1 x_2(t) + c_1 x_1(t-\tau)\exp(-x_1(t-\tau))\right], \\ \frac{dx_2}{dt} &= r(t)\left[-a_2 x_2(t) + b_2 x_1(t) + c_2 x_2(t-\tau)\exp(-x_2(t-\tau))\right], \end{aligned} \qquad (9.1)$$

where $a_i, b_i, c_i, \tau = \text{const} \geq 0$, and $r(t), t \geq 0$, is a piece-wise continuous function bounded on the positive half-line with the property

$$\inf_{t \geq 0} r(t) > 0.$$

Take the initial conditions

$$x_1(t) = \phi_1(t), \quad x_2(t) = \phi_2(t) \ \ (t \in [-\tau, 0]) \tag{9.2}$$

with continuous functions ϕ_k. Everywhere in this section it is assumed that ϕ_k $(k = 1, 2)$ are non-negative.

Recall that inequalities for vector-valued functions are understood in the coordinate-wise sense.

Rewrite (9.1) as

$$\frac{dx(t)}{dt} = r(t)(Ax(t) + CF_1(x(t-\tau))), \tag{9.3}$$

where

$$A = \begin{pmatrix} -a_1 & b_1 \\ b_2 & -a_2 \end{pmatrix}, \quad C = \begin{pmatrix} c_1 & 0 \\ 0 & c_2 \end{pmatrix}$$

and

$$F_1(x(t-\tau)) = (x_k(t-\tau)\exp(-x_k(t-\tau)))_{k=1}^2.$$

Taking $Fx(t) = CF_1(x(t-\tau)$ we have

$$0 \leq Fv(t) \leq Cv(t-\tau) \ (v(t) \geq 0).$$

Equation (8.7) in the considered case takes the form

$$\frac{dy(t)}{dt} = r(t)[Ay(t) + Cy(t-\tau)], \tag{9.4}$$

and equation (8.8) takes the form

$$\frac{dy(t)}{dt} = r(t)Ay(t). \tag{9.5}$$

The evolution operator $U_-(t, s)$ to (9.5) is

$$U_-(t, \tau) = e^{\int_\tau^t r(s)ds \, A}.$$

Since the off-diagonal entries of A are non-negative, the evolution operator to (9.5) is non-negative. Applying Theorem 10.8.2 and omitting simple calculations, we arrive at the following result.

Corollary 10.9.1. *Any solution $x(t)$ of problem (9.1), (9.2) is non-negative and satisfies the inequalities*

$$e^{\int_0^t r(s)ds\, A}\phi(0) \le x(t) \le x_+(t) \ \ (t \ge 0)$$

where $x_+(t)$ is the solution of problem (9.4), (9.2). Consequently, the zero solution to (9.1) is globally asymptotically stable with respect to the cone $K = C([-\eta\,,0], R^2_+)$, provided (9.4) is asymptotically stable.

Rewrite (9.4) as

$$y(t) = U_-(t,0)\phi(0) + \int_0^t U_-(t,s)r(s)Cy(s-\tau)ds.$$

Thus

$$\|y\|_{C(0,T)} = \sup_t \|U_-(t,0)\phi(0)\|_2 + \|y\|_{C(-\tau,T)}M_0 \ \ (0 < T < \infty),$$

where

$$M_0 = \sup_t \int_0^t r(s)\|U_-(t,s)C\|_2 ds.$$

Here $\|.\|_2$ is the spectral norm in $\mathbb{C}^2$. Hence it easily follows that (9.4) is stable provided $M_0 < 1$. Note that the eigenvalues $\lambda_1(A)$ and $\lambda_2(A)$ of A are simply calculated and with the notation

$$m(t,s) = \int_s^t r(s_1)ds_1,$$

we have

$$\|e^{m(t,s)\, A}\|_2 \le e^{\alpha(A)m(t,s)}(1 + g(A)m(t,s)),$$

where $\alpha(A) = \max_{k=1,2} \operatorname{Re}\lambda_1(A)$ and $g(A) = |b_1 - b_2|$ (see Section 2.8). Assume that $c_1 \le c_2$, $\alpha(A) < 0$, and

$$r_- = \inf_t r(t) > 0, \quad \text{and} \quad r_+ = \sup_t r(t) < \infty.$$

Then $M_0 \le M_1$, where

$$M_1 := r_+ c_2 \sup_t \int_0^t e^{(t-s)\alpha(A)r_-}(1 + g(A)(t-s)r_+\,)ds$$
$$= c_2 r_+ \int_0^\infty e^{u\alpha(A)r_-}(1 + g(A)ur_+)du.$$

This integral is simply calculated. Thus, if $M_1 < 1$, then the zero solution to (9.1) is globally asymptotically stable with respect to the cone $C([-\eta\,,0], R^2_+)$.

10.10 Input-to-state stability of general systems

This section is devoted to the input-to-state stability of coupled systems of functional differential equations with nonlinear causal mappings. The notion of the input-to-state stability plays an essential role in control theory.

Consider in $\mathbb{C}^n$ the problem

$$\dot{x}(t) = \int_0^\eta dR_0(\tau)x(t-\tau) + [Fx](t) + u(t) \quad (0 < \eta < \infty;\ t > 0), \tag{10.1}$$

$$x(t) = 0 \quad \text{for} \quad -\eta \le t \le 0, \tag{10.2}$$

where $u(t)$ is the input, $x(t)$ is the state, and $R_0(\tau)$ $(0 \le \tau \le \eta)$ is an $n \times n$-matrix-valued function of bounded variation, again. In addition, F is a continuous causal mapping in $L^p(-\eta, \infty)$ $(p \ge 1)$.

For example, we can reduce the form (10.1) to the nonlinear differential delay equation

$$\dot{x}(t) = \int_0^\eta dR_0(\tau)x(t-\tau) + F_0(x(t-h)) + u(t) \quad (h = \text{const} > 0; t \ge 0), \tag{10.3}$$

where $F_0 : \mathbb{C}^n \to \mathbb{C}^n$ is a continuous function, $u \in L^p(0,\infty)$. Equation (10.3) takes the form (10.1) with

$$[Fx](t) = F_0(x(t-h)).$$

Besides F is causal in $L^\infty(-h,\infty)$.

Again,

$$K(z) = zI - \int_0^\eta e^{-\tau z} dR_0(\tau) \quad \text{and} \quad G(t) = \frac{1}{2\pi}\int_{-\infty}^{\infty} \frac{e^{i\omega t} d\omega}{K(i\omega)}$$

are the characteristic matrix-valued function of and fundamental solution of the linear equation

$$\dot{z} = \int_0^\eta dR_0(\tau)z(t-\tau) \quad (t \ge 0), \tag{10.4}$$

respectively. All the characteristic values of $K(.)$ are in C_-.

Definition 10.10.1. We will say that system (10.1), (10.2) is input-to-state L^p-stable, if there is a positive constant $m_0 \le \infty$, such that for all $u \in L^p$ satisfying $\|u\|_{L^p(0,\infty)} \le m_0$, a solution of problem (10.1), (10.2) is in $L^p(0,\infty)$.

Put

$$\Omega_p(\varrho) = \{w \in L^p(-\eta,\infty) : \|w\|_{L^p(-\eta,\infty)} \le \varrho\}$$

for a positive number $\varrho \le \infty$.

It is assumed that there is a constant $q = q(\varrho) \ge 0$, such that

$$\|Fw\|_{L^p(0,\infty)} \le q\|w\|_{L^p(-\eta,\infty)} \quad (w \in \Omega_p(\varrho)). \tag{10.6}$$

Recall that

$$\hat{G}f(t) = \int_0^t G(t,s)f(s)ds.$$

Theorem 10.10.2. *Let the conditions* (10.6) *and* $q\|\hat{G}\|_{L^p(0,\infty)} < 1$ *hold. Then* (10.1) *is input-to-state* L^p*-stable.*

Proof. Put $l = \|u\|_{L^p(0,\infty)}$ for a fixed u. Repeating the arguments of the proof of Theorem 10.1.2 with L^p instead of C, and taking into account the zero initial conditions, we get the inequality

$$\|x\|_{L^p(0,\infty)} \leq \frac{\|\hat{G}\|_{L^p(0,\infty)} l}{1 - q\|\hat{G}\|_{L^p(0,\infty)}}.$$

This inequality proves the theorem. □

By the properties of convolutions (see Section 1.3), we have

$$\|\hat{G}\|_{L^p(),\infty)} \leq \|G\|_{L^1(0,\infty)}.$$

Now Theorem 10.10.2 implies our next result.

Corollary 10.10.3. *Let condition* (10.6) *hold. In addition, let* $q\|G\|_{L^1(0,\infty)} < 1$. *Then equation* (10.1) *is input-to-state* L^p*-stable.*

Due to Lemma 4.4.1 we have

$$\|\hat{G}\|_{L^2(0,\infty)} \leq \theta(K), \quad \text{where} \quad \theta(K) := \sup_{-2\ \mathrm{var}(R_0) \leq \omega \leq 2\ \mathrm{var}(R_0)} \|K^{-1}(i\omega)\|_n.$$

Now making use of Theorem 10.10.2, we arrive at the following result.

Corollary 10.10.4. *Let condition* (10.6) *hold with* $p = 2$. *In addition, let* $q\theta(K) < 1$. *Then system* (10.1) *is input-to-state* L^2*-stable.*

10.11 Input-to-state stability of systems with one delay in linear parts

In this section we illustrate the results of the previous section in the case of systems whose linear parts have one distributed delay:

$$\dot{x}(t) + A\int_0^\eta x(t-s)d\mu(s) = [Fx](t) + u(t), \tag{11.1}$$

where μ is a scalar nondecreasing function and A is a constant positive definite Hermitian matrix. So the eigenvalues $\lambda_j(A)$ of A are real and positive. Then the linear equation

$$\dot{x}(t) + A\int_0^\eta x(t-s)d\mu(s) = 0 \tag{11.2}$$

can be written as

$$\dot{x}_j(t) - \lambda_j(A)E_\mu x_j(t) = 0, \ \ j = 1, \ldots, n, \tag{11.3}$$

where

$$E_\mu f(t) = \int_0^\eta f(t-s)d\mu(s)$$

for a scalar function f.

If the inequality

$$e\eta\, \lambda_j(A) \ \mathrm{var}(\mu) < 1 \ \ (j = 1, \ldots, n) \tag{11.4}$$

holds, then by Lemma 4.6.6 the fundamental solution G_j of the scalar equation (11.3) is positive and

$$\int_0^\infty G_j(t)dt = \frac{1}{\lambda_j(A)\,\mathrm{var}(\mu)}.$$

But the fundamental solution G_μ of the vector equation (11.2) satisfies the equality

$$\int_0^\infty \|G_\mu(t)\|_n dt = \max_j \int_0^\infty G_j(t)dt.$$

Thus

$$\int_0^\infty \|G_\mu(t)\|_n dt = \frac{1}{\min_j \lambda_j(A)\,\mathrm{var}(\mu)}.$$

Now Theorem 10.10.2 implies our next result.

Corollary 10.11.1. *Let A be a positive definite Hermitian matrix. and conditions* (10.6) *and* (11.4) *hold. If in addition,*

$$q < \min_j \lambda_j(A)\,\mathrm{var}(\mu), \tag{11.5}$$

then system (11.1) *is input-to-state L^p-stable.*

10.12 Comments

This chapter is particularly based on the papers [39, 41] and [61].

The basic results on the stability of nonlinear differential-delay equations are presented, in particular, in the well-known books [72, 77, 100]. About the recent results on absolute stability of nonlinear retarded systems see [88, 111] and references therein.

The stability theory of nonlinear equations with causal mappings is at an early stage of development. The basic method for stability analysis is the direct Lyapunov method, cf. [13, 83]. But finding Lyapunov functionals for equations with causal mappings is a difficult mathematical problem.

Interesting investigations of linear causal operators are presented in the books [82, 105]. The papers [5, 15] also should be mentioned. In the paper [15], the existence and uniqueness of local and global solutions to the Cauchy problem for equations with causal operators in a Banach space are established. In the paper [5] it is proved that the input-output stability of vector equations with causal operators is equivalent to the causal invertibility of causal operators.

Chapter 11

Scalar Nonlinear Equations

In this chapter, nonlinear scalar first- and higher-order equations with differential-delay linear parts and nonlinear causal mappings are considered. Explicit stability conditions are derived.

The Aizerman–Myshkis problem is also discussed.

11.1 Preliminary results

In this chapter all the considered functions are scalar; so $L^p(a,b) = L^p([a,b],\mathbb{C})$ and $C(a,b) = C([a,b],\mathbb{C})$. In addition,

$$\textit{either} \quad X(a,b) = L^p(a,b), p \geq 1, \quad \textit{or} \quad X(a,b) = C(a,b).$$

Consider the equation

$$x(t) = z(t) + \int_0^t k(t,t_1)(Fx)(t_1)dt_1 \ \ (t \geq 0), \tag{1.1a}$$

$$x(t) = z(t) \ \ (-\eta \leq t \leq 0), \tag{1.1b}$$

where $k : [0 \leq s \leq t \leq \infty) \to \mathbb{R}$ is a measurable kernel and $z \in X(-\eta, \infty)$ is given, and F is a continuous causal mapping in $X(-\eta\,, \infty)$ $(0 \leq \eta < \infty)$ (see Section 1.8). For instance, the mapping defined by

$$(Fw)(t) = \int_0^\eta g(w(t-s))d\mu(s) \quad (t \geq -\eta\,), \tag{1.2}$$

where μ is a nondecreasing function and g is a continuous function with $g(0) = 0$, is causal in $C(-\eta\,, \infty)$.

A solution of (1.1) is a continuous function x defined on $[-\eta\,, \infty)$, which satisfies (1.1).

It is assumed that there is a constant $q \geq 0$, such that

$$\|Fv\|_{X(0,\infty)} \leq q\, \|v\|_{X(-\eta\,,\infty)} \ (v \in X(-\eta\,, \infty)). \tag{1.3}$$

Introduce the operator $V : X(0,\infty) \to X(0,\infty)$ by

$$(Vv)(t) = \int_0^t k(t,t_1)v(t_1)dt_1 \ \ (t > 0;\ v \in X(0,\infty)).$$

Lemma 11.1.1. *Let V be compact in $X(0,\tau)$ for each finite τ, and the conditions* (1.3) *and*

$$q\|V\|_{X(0,\infty)} < 1 \tag{1.4}$$

hold. Then equation (1.1) *has a (continuous) solution $x \in X(-\eta\,,\infty)$. Moreover, that solution satisfies the inequality*

$$\|x\|_{X(-\eta\,,\infty)} \le \frac{\|z\|_{X(-\eta\,,\infty)}}{1 - q\|V\|_{X(0,\infty)}}.$$

Proof. By Lemmas 10.1.1 and 10.1.3, if condition (1.3) holds, then for all $T \ge 0$ and $w \in X(-\eta\,,T)$, we have

$$\|Fw\|_{X(0,T)} \le q\ \|w\|_{X(-\eta\,,T)}.$$

On $X(-\eta\,,T)$, $T < \infty$, let us define the mapping Φ by

$$(\Phi w)(t) = z(t) + (VFw)(t), t \ge 0, \quad \text{and} \quad (\Phi w)(t) = z(t), t < 0,$$

for a $w \in X(-\eta\,,T)$. Hence, according to the previous inequality, for any number $r > 0$, large enough, we have

$$\|\Phi w\|_{X(-\eta\,,T)} \le \|z\|_{X(-\eta\,,T)} + \|V\|_{X(0,T)}q\|w\|_{X(-\eta\,,T)} \le r \ \ (\|w\|_{X(-\eta\,,T)} \le r).$$

So Φ maps a bounded set of $X(-\eta\,,T)$ into itself. Now the existence of a solution $x(t)$ is due to the Schauder Fixed Point Theorem, since V is compact.

From (1.1) it follows that

$$\|x\|_{X(-\eta\,,T)} \le \|z\|_{X(-\eta\,,T)} + \|V\|_{X(0,T)}q\|x\|_{X(-\eta\,,T)}.$$

Thus (1.4) implies

$$\|x\|_{X(-\eta\,,T)} \le \frac{\|z\|_{X(-\eta\,,T)}}{1 - q\|V\|_{X(0,T)}}.$$

Now letting $T \to \infty$ we get the required result. □

Furthermore, let $k(t,s) = Q(t-s)$ with a continuous $Q \in L^1(0,\infty)$. Then $V = V_Q$, where the operator V_Q is defined by

$$(V_Qv)(t) = \int_0^t Q(t-t_1)v(t_1)dt_1 \ \ (v \in X(0,\infty);\ t > 0).$$

For each positive $T < \infty$, operator V_Q is compact. Moreover, by the properties of the convolution operators we have

$$\|V_Q\|_{X(0,\infty)} \leq \|Q\|_{L^1(0,\infty)}$$

with $X(0,\infty) = L^p(0,\infty)$ and $X(0,\infty) = C(0,\infty)$ (see Section 1.3). Now the previous lemma implies

Corollary 11.1.2. *Assume that $Q \in L^1(0,\infty)$, $z \in X(-\eta,\infty)$, and the condition* (1.3) *holds. If, in addition, $q\|Q\|_{L^1(0,\infty)} < 1$, then the problem*

$$x(t) = z(t) + \int_0^t Q(t-t_1)(Fx)(t_1)dt_1 \quad (t > 0), \tag{1.5a}$$

$$x(t) = z(t) \quad (-\eta \leq t \leq 0) \tag{1.5b}$$

has a solution $x \in X(-\eta,\infty)$. Moreover, that solution satisfies the inequality

$$\|x\|_{X(-\eta,\infty)} \leq \frac{\|z\|_{X(-\eta,\infty)}}{1 - q\|Q\|_{L^1(0,\infty)}}.$$

Let

$$\tilde{Q}(z) := \int_0^\infty e^{-zt}Q(t)dt \quad (\operatorname{Re} z \geq 0)$$

be the Laplace transform of Q. Then by the Parseval equality we easily get $\|V_Q\|_{L^2(0,\infty)} = \Lambda_Q$, where

$$\Lambda_Q := \sup_{s\in\mathbb{R}} \|\tilde{Q}(is)\|.$$

So Lemma 11.1.1 implies our next result.

Corollary 11.1.3. *Assume that $Q \in L^1(-\eta,\infty)$, $z \in L^2(-\eta,\infty)$, and the condition* (1.3) *holds with $X(-\eta,\infty) = L^2(-\eta,\infty)$. If, in addition, the condition $q\Lambda_Q < 1$ is fulfilled, then problem* (1.5) *has a (continuous) solution $x \in L^2(-\eta,\infty)$. Moreover, that solution satisfies the inequality*

$$\|x\|_{L^2(-\eta,\infty)} \leq \frac{\|z\|_{L^2(-\eta,\infty)}}{1 - q\Lambda_Q}.$$

Furthermore, suppose that

$$Q(t) \geq 0 \ (t \geq 0). \tag{1.6}$$

Then, obviously,

$$|\tilde{Q}(iy)| = \left|\int_0^\infty e^{-yit}Q(t)dt\right| \leq \int_0^\infty Q(t)dt = \tilde{Q}(0) \quad (y \in \mathbb{R}). \tag{1.7}$$

Now Corollary 11.1.2 implies the following result.

Corollary 11.1.4. *Let $Q \in L^1(0,\infty), z \in X(-\eta,\infty)$ and the conditions* (1.3), (1.6) *and*

$$q\tilde{Q}(0) < 1 \tag{1.8}$$

be fulfilled. Then problem (1.5) *has a solution $x \in X(-\eta,\infty)$ and*

$$\|x\|_{X(-\eta,\infty)} \le \frac{\|z\|_{X(-\eta,\infty)}}{1 - q\tilde{Q}(0)}.$$

To investigate the stability of the scalar differential delay equation we need the following lemma.

Lemma 11.1.5. *Let $Q \in L^1(0,\infty)$ and condition* (1.6) *hold. Then inequality* (1.8) *is valid if and only if all the zeros of the function*

$$\frac{1}{\tilde{Q}(z)} - q \tag{1.9}$$

are in $C_- := \{z \in \mathbb{C} : \operatorname{Re} z < 0\}$.

Proof. Let (1.8) hold. Then thanks to (1.7), $\frac{1}{|\tilde{Q}(iy)|} > q$ for all real y. So according to the Rouché theorem, all the roots of the function defined by (1.9) are in C_-, since $Q \in L^1(0,\infty)$ and therefore all the roots of the function $1/\tilde{Q}(z)$ are in C_-.

Conversely, let all the roots of the function defined by (1.9) be in C_-. Then either

$$\frac{1}{|\tilde{Q}(iy)|} > q \tag{1.10}$$

or

$$\frac{1}{|\tilde{Q}(iy)|} < q \tag{1.11}$$

for all real y. But in the case (1.11), the integral

$$\int_{-\infty}^{\infty} \tilde{Q}(iy)dy$$

does not converge. This proves the lemma. □

11.2 Absolute stability

In the rest of this chapter the uniqueness of solutions is assumed.

Let us consider the equation

$$x^{(n)}(t) + \sum_{k=0}^{n-1} \int_0^{\eta} x^{(k)}(t-\tau)d\mu_k(\tau) = [Fx](t) \quad (t > 0), \tag{2.1}$$

where μ_k $(k = 0, \ldots, n-1)$ are bounded nondecreasing functions defined on $[0, \eta\,]$ and F is a causal mapping in $X(-\eta\,, \infty)$ with $X(a,b) = L^p(a,b), p \geq 1$, or $X(a,b) = C(a,b)$, again. Impose the initial condition

$$x(t) = \phi(t) \; (-\eta \; \leq t \leq 0) \tag{2.2}$$

with a given function ϕ having continuous derivatives up to $n-1$th order. Let $K(.)$ be the characteristic function of the equation

$$x^{(n)}(t) + \sum_{k=0}^{n-1} \int_0^\eta x^{(k)}(t-\tau)d\mu_k(\tau) = 0 \;\; (t > 0). \tag{2.3}$$

That is,

$$K(\lambda) = \lambda^n + \sum_{k=0}^{n-1} \lambda^k \int_0^\eta e^{-\lambda\tau} d\mu_k(\tau).$$

It is assumed that all the zeros of $K(.)$ are in C_-. Introduce the Green function of (2.3):

$$G(t) := \frac{1}{2\pi} \int_{-\infty}^{\infty} \frac{e^{ti\omega} d\omega}{K(i\omega)} \;\; (t \geq 0).$$

If $n = 1$, then the notions of the Green function and fundamental solution coincide.

It is simple to check that the equation

$$w^{(n)}(t) + \sum_{k=0}^{n-1} \int_0^\eta w^{(k)}(t-\tau)d\mu_k(\tau) = f(t) \;\; (t \geq 0) \tag{2.4}$$

with the zero initial condition

$$w^{(k)}(t) \equiv 0 \; (-\eta \; \leq t \leq 0; \; k = 1, \ldots, n-1) \tag{2.5}$$

and the locally integrable function f satisfying

$$|f(t)| \leq c_0 e^{c_1 t} \;\; (c_0, c_1 = \text{const}; \; t \geq 0)$$

admits the Laplace transform. So by the inverse Laplace transform, problem (2.4), (2.5) has the solution

$$w(t) = \int_0^t G(t-t_1)f(t_1)dt_1 \;\; (t > 0).$$

Hence it follows that (2.1) is equivalent to the equation

$$x(t) = \zeta(t) + \int_0^t G(t-t_1)(Fx)(t_1)dt_1 \;\; (t > 0), \tag{2.6}$$

where ζ is a solution of problem (2.3), (2.2).

A continuous solution of the integral equation (2.6) with condition (2.2) will be called *a (mild) solution of problem* (2.1), (2.2).

Lemma 11.2.1. *Let all the zeros of $K(z)$ be in C_-. Then the linear equation* (2.3) *is exponentially stable.*

Proof. As it is well known, if all the zeros of $K(z)$ are in C_-, then (2.3) is asymptotically stable, cf. [77, 78]. Now we get the required result by small perturbations and the continuity of the zeros of $K(z)$. □

Introduce in $X(0,\infty)$ the operator

$$\hat{G}w(t) = \int_0^t G(t-t_1)w(t_1)dt_1 \quad (t>0).$$

Thanks to the preceding lemma it is bounded, since

$$\|\hat{G}\|_{X(0,\infty)} \le \|G\|_{L^1(0,\infty)} < \infty.$$

Moreover, since (2.3) is exponentially stable, for a solution ζ of problem (2.2), (2.3) we have

$$|\zeta(t)| \le \text{const}\ e^{-\epsilon t}\sum_{k=0}^{n-1} \|\phi^{(k)}\|_{C[-\eta\,,0]} \ (\epsilon>0;\ t\ge 0). \tag{2.7}$$

Hence, $\zeta \in X(-\eta\,,\infty)$. Now Lemma 11.1.1 implies the following result.

Theorem 11.2.2. *Assume that condition* (1.3) *holds for $X(-\eta\,,\infty) = L^p(-\eta\,,\infty)$, $p \ge 1$ or $X(-\eta\,,\infty) = C(-\eta\,,\infty)$, and all the zeros of $K(z)$ are in C_-. If, in addition,*

$$q\|\hat{G}\|_{X(0,\infty)} < 1, \tag{2.8}$$

then problem (2.1), (2.2) *has a solution $x \in X(-\eta\,,\infty)$ and*

$$\|x\|_{X(0,\infty)} \le \frac{\|\zeta\|_{X(-\eta\,,\infty)}}{1-q\|\hat{G}\|_{X(0,\infty)}},$$

where ζ is a solution of problem (2.2), (2.3), *and consequently,*

$$\|x\|_{X(0,\infty)} \le M\sum_{k=0}^{n-1} \|\phi^{(k)}\|_{C[-\eta\,,0]}, \tag{2.9}$$

where the constant M does not depend on the initial conditions.

Combining this theorem with Corollary 11.1.4 we obtain the following result.

Corollary 11.2.3. *Assume that condition* (1.3) *holds with $X(-\eta,\infty) = L^p(-\eta\,,\infty)$ for a $p \ge 1$, or $X(-\eta\,,\infty) = C(-\eta\,,\infty)$, and all the zeros of $K(z)$ are in C_-. If, in addition, $G(t)$ is non-negative and*

$$K(0) > q, \tag{2.10}$$

then problem (2.1), (2.2) *has a solution $x \in X(0,\infty)$. Moreover, that solution satisfies inequality* (2.9).

Furthermore, Corollary 11.1.3 implies

Corollary 11.2.4. *Let all the zeros of $K(z)$ be in C_-. Assume that the conditions* (1.3) *with $X(-\eta\,,\infty) = L^2(-\eta\,,\infty)$ and*

$$\inf_{s\in\mathbb{R}} |K(is)| > q \tag{2.11}$$

hold. Then problem (2.1), (2.2) *has a solution $x \in L^2(0,\infty)$. Moreover, that solution satisfies the inequality* (2.9) *with $X(0,\infty) = L^2(0,\infty)$.*

Definition 11.2.5. Equation (2.1) is said to be absolutely X-stable in the class of nonlinearities satisfying (1.3), if there is a positive constant M_0 independent of the specific form of mapping F (but dependent on q), such that (2.9) holds for any solution $x(t)$ of (2.1).

Let us appeal to the following corollary to Theorem 11.2.2.

Corollary 11.2.6. *Assume that all the zeros of $K(z)$ be in C_-. and condition* (2.8) *holds, then* (2.1) *is absolutely X-stable in the class of nonlinearities* (1.3) *with $X(-\eta\,,\infty) = L^p(-\eta\,,\infty), p \geq 1$ or $X(-\eta\,,\infty) = C(-\eta\,,\infty)$.*

If, in addition, G is positive, then condition (2.8) *can be replaced by* (2.10).

In the case $X(-\eta,\infty) = L^2(-\eta,\infty)$, condition (2.8) *can be replaced by* (2.11).

11.3 The Aizerman–Myshkis problem

We consider the following problem, which we call the Aizerman–Myshkis problem.

Problem 11.3.1: *To separate a class of equations* (2.1), *such that the asymptotic stability of the linear equation*

$$x^{(n)} + \sum_{k=0}^{n-1} \int_0^{\eta} x^{(k)}(t-\tau)d\mu_j(\tau) = \tilde{q}x(t), \tag{3.1}$$

with some $\tilde{q} \in [0,q]$ provides the absolute X-stability of (2.1) *in the class of nonlinearities* (1.3).

Recall that $X(-\eta\,,\infty) = C(-\eta\,,\infty)$ or $X(-\eta\,,\infty) = L^p(-\eta\,,\infty)$.

Theorem 11.3.1. *Let the Green function of* (2.3) *be non-negative and condition* (2.10) *hold. Then equation* (2.1) *is absolutely L^2-stable in the class of nonlinearities satisfying* (1.3). *Moreover,* (2.1) *satisfies the Aizerman–Myshkis problem in $L^2(-\eta\,,\infty)$ with $\tilde{q} = q$.*

Proof. Corollary 11.2.6 at once yields that (2.1) is absolutely stable, provided (2.10) holds. By Lemma 11.1.5 this is equivalent to the asymptotic stability of (3.1) with $\tilde{q} = q$. This proves the theorem. □

Let us consider the first-order equation

$$\dot{x}(t) + \int_0^\eta x(t-s)d\mu(s) = [Fx](t) \ \ (t \geq 0), \tag{3.2}$$

where μ is a nondecreasing function having a bounded variation $\mathrm{var}(\mu)$. We need the corresponding linear equation

$$\dot{y}(t) + \int_0^\eta y(t-s)d\mu(s) = 0 \ \ (t > 0). \tag{3.3}$$

Denote by $G_\mu(t)$ the Green function (the fundamental solution) of this equation. In the next section we prove the following result.

Lemma 11.3.2. *Under the condition,*

$$e\eta \ \ \mathrm{var}(\mu) < 1, \tag{3.4}$$

G_μ *is non-negative on the positive half-line.*

Denote by K_μ the characteristic function of (3.3):

$$K_\mu(z) = z + \int_0^\eta e^{-sz} d\mu(s).$$

So $K_\mu(0)$ is equal to the variation $\mathrm{var}(\mu)$ of μ. By Lemma 11.1.5, $K_\mu(z) - q$ has all its zeros in C_- if and only if $q < \mathrm{var}(\mu)$. Now by Theorem 11.3.1 we easily get the following result.

Corollary 11.3.3. *Assume that the conditions* (3.4) *and* $q < \mathrm{var}(\mu)$ *are fulfilled. Then equation* (3.2) *is* X*-absolutely stable in the class of nonlinearities* (1.3).

Now let us consider the second-order equation

$$\ddot{u}(t) + A\dot{u}(t) + B\dot{u}(t-1) + Cu(t) + Du(t-1) + Eu(t-2) = [Fu](t) \ \ (t > 0) \tag{3.5}$$

with non-negative constants A, B, C, D, E.

Introduce the functions

$$K_2(\lambda) = \lambda^2 + A\lambda + B\lambda e^{-\lambda} + C + De^{-\lambda} + Ee^{-2\lambda} \tag{3.6}$$

and

$$G_2(t) := \frac{1}{2\pi i} \int_{c_0 - \infty i}^{c_0 + i\infty} \frac{e^{tz} dz}{K_2(z)} \ \ (c_0 = \mathrm{const}).$$

Assume that

$$B^2/4 > E, \ A^2/4 > C, \tag{3.7}$$

and let

$$r_\pm(A, C) = \frac{A}{2} \pm \sqrt{\frac{A^2}{4} - C}, \ \ \text{and} \ \ r_\pm(B, E) = \frac{B}{2} \pm \sqrt{\frac{B^2}{4} - E}.$$

In the next section we also prove the following result.

Lemma 11.3.4. *Let the conditions* (3.7),

$$D \le r_+(B,E)r_-(A,C) + r_-(B,E)r_+(A,C) \tag{3.8}$$

and

$$r_+(B,E)e^{r_+(A,C)} < \frac{1}{e} \tag{3.9}$$

hold. Then $G_2(t)$ is non-negative on the positive half-line.

Clearly, $K_2(0) = C + D + E$. Now Theorem 11.3.1 and Lemma 11.1.5 yield the following corollary.

Corollary 11.3.5. *Let the conditions* (3.7)–(3.9) *and $q < C + D + E$ hold. Then equation* (3.5) *is X-absolutely stable in the class of nonlinearities* (1.3).

Finally, consider the higher-order equations. To this end at a continuous function v defined on $[-\eta\,,\infty)$, let us define an operator S_k by

$$(S_k v)(t) = a_k v(t - h_k) \ \ (k = 1,\dots,n;\ a_k = \text{const} > 0, h_k = \text{const} \ge 0;\ t \ge 0).$$

Besides,

$$h_1 + \dots + h_n = \eta\,.$$

Consider the equation

$$\prod_{k=1}^{n}\left(\frac{d}{dt} + S_k\right)x(t) = [Fx](t) \ \ (t \ge 0). \tag{3.10}$$

Put

$$\hat{W}_n(z) := \prod_{k=1}^{n}(z + a_k e^{-h_k z}) \quad \text{and} \quad G_n(t) = \frac{1}{2\pi i}\int_{c_0 - i\infty}^{c_0 + i\infty}\frac{e^{zt}dz}{\hat{W}_n(z)}.$$

So $G_n(t)$ is the Green function of the linear equation

$$\prod_{k=1}^{n}\left(\frac{d}{dt} + S_k\right)x(t) = 0.$$

Due to the properties of the convolution, we have

$$G_n(t) = \int_0^t \hat{G}_1(t - t_1)\int_0^{t_1}\hat{G}_2(t_1 - t_2)\dots\int_0^{t_{n-2}}\hat{G}_n(t_{n-1})dt_{n-1}\,\dots\,dt_2\,dt_1,$$

where $\hat{G}_k(t)$ $(k = 1,\dots,n)$ is the Green function of the equation

$$\left(\frac{d}{dt} + S_k\right)x(t) = \dot{x}(t) + a_k x(t - h_k) = 0.$$

Assume that

$$ea_k h_k < 1 \quad (k = 1, \dots, n); \tag{3.11}$$

then due to Lemma 11.3.2 $\hat{G}_k(t) \geq 0$ $(t \geq 0;\ k = 1, \dots, n)$. We thus have proved the following result.

Lemma 11.3.6. *Let condition* (3.11) *hold. Then* $G_n(t)$ *is non-negative.*

Clearly,

$$\hat{W}_n(0) = \prod_{k=1}^{n} a_k \,.$$

Now Theorem 11.3.1 and Lemma 11.1.5 yield our next result.

Corollary 11.3.7. *Let the conditions* (3.11) *and*

$$q < \prod_{k=1}^{n} a_k$$

hold. Then equation (3.10) *is* X*-absolutely stable in the class of nonlinearities* (1.3).

11.4 Proofs of Lemmas 11.3.2 and 11.3.4

With constants $a \geq 0,\ b > 0$, let us consider the equation

$$\dot{u}(t) + au(t) + bu(t-h) = 0. \tag{4.1}$$

Lemma 11.4.1. *Let the condition*

$$hbe^{ah} < e^{-1} \tag{4.2}$$

hold. Then the Green function of equation (4.1) *is positive.*

Proof. First, we consider the Green function $G_b(t)$ of the equation

$$\dot{u} + bu(t-h) = 0 \ (b = \text{const},\ t \geq 0). \tag{4.3}$$

Recall that G_b satisfies with the initial condition

$$G_b(0) = 1,\ G_b(t) = 0 \ (t < 0). \tag{4.4}$$

Suppose that

$$bh < e^{-1}. \tag{4.5}$$

Since

$$\max_{\tau \geq 0} \tau e^{-\tau} = e^{-1},$$

there is a positive solution w_0 of the equation $we^{-w} = bh$. Taking $c = h^{-1}w_0$, we get a solution c of the equation $c = be^{hc}$. Put in (4.3) $G_b(t) = e^{-ct}z(t)$. Then

$$\dot{z} - cz + be^{hc}z(t-h) = \dot{z} + c(z(t-h) - z(t)) = 0.$$

But due to (4.4) $z(0) = 1, z(t) = 0$ $(t < 0)$. So the latter equation is equivalent to the following one:

$$z(t) = 1 + \int_0^t c[z(s) - z(s-h)]ds = 1 + c\int_0^t z(s)ds - c\int_0^{t-h} z(s)ds.$$

Consequently,

$$z(t) = 1 + c\int_{t-h}^t z(s)ds.$$

Due to the von Neumann series it follows that $z(t)$ and, therefore, the Green function $G_b(t)$ of (4.3) are positive.

Furthermore, substituting $u(t) = e^{-at}v(t)$ into (4.1), we have the equation

$$\dot{v}(t) + be^{ah}v(t-h) = 0.$$

According to (4.5), condition (4.2) provides the positivity of the Green function of the latter equation. Hence the required result follows. □

Denote by G_+ the Green function of equation (4.3) with $b = \text{var}(\mu)$. Due to the previous lemma, under condition (3.4) G_+ is positive.

The assertion of Lemma 11.3.2 follows from the next result.

Lemma 11.4.2. *If condition* (3.4) *holds, then the Green function* G_μ *of equation* (3.3) *is non-negative and satisfies the inequality*

$$G_\mu(t) \geq G_+(t) \geq 0 \ (t \geq 0).$$

Proof. Indeed, according to the initial conditions, for a sufficiently small $t_0 > \eta$,

$$G_\mu(t) \geq 0, \ \dot{G}_\mu(t) \leq 0 \ (0 \leq t \leq t_0).$$

Thus,

$$G_\mu(t-\eta) \geq G_\mu(t-s) \ (s \leq \eta; \ 0 \leq t \leq t_0).$$

Hence,

$$\text{var}\,(\mu)G_\mu(t-\eta) \geq \int_0^\eta G_\mu(t-s)d\mu(s) \ (t \leq t_0).$$

According to (3.3) we get

$$\dot{G}_\mu(t) + \text{var}\,(\mu)G_\mu(t-\eta) = f(t)$$

with

$$f(t) = \text{var}\,(\mu)G_\mu(t-\eta\,) - \int_0^\eta G_\mu(t-s)d\mu(s) \ge 0 \ (0 \le t \le t_0).$$

Hence, by virtue of the variation of constants formula, arrive at the relation

$$G_\mu(t) = G_+(t) + \int_0^t G_+(t-s)f(s)ds \ge G_+(t) \ (0 \le t \le t_0).$$

Extending this inequality to the whole half-line, we get the required result. □

Proof of Lemma 11.3.4. First, consider the function

$$K_0(\lambda) = (\lambda + a_1 + b_1 e^{-\lambda})(\lambda + a_2 + b_2 e^{-\lambda})$$

with non-negative constants a_k, b_k $(k = 1, 2)$. Then

$$G_0(t) := \frac{1}{2\pi i}\int_{c_0 - \infty i}^{c_0 + i\infty} \frac{e^{tz}dz}{K_0(z)} \quad (c_0 = \text{const})$$

is the Green function to equation

$$\left(\frac{d}{dt} + S_1\right)\left(\frac{d}{dt} + S_2\right)x(t) = 0 \ \ (t \ge 0) \tag{4.6}$$

where

$$(S_k v)(t) = a_k v(t) + b_k v(t-1) \ \ (k = 1, 2;\ t \ge 0).$$

Due to the properties of the convolution, we have

$$G_0(t) = \int_0^t W_1(t - t_1)W_2(t_1)dt_1,$$

where $W_k(t)$ $(k = 1, 2)$ is the Green function of the equation

$$\left(\frac{d}{dt} + S_k\right)x(t) = \dot{x}(t) + a_k x(t) + b_k x(t-1) = 0.$$

Assume that

$$e^{a_k} b_k < \frac{1}{e} \quad (k = 1, 2), \tag{4.7}$$

then due to Lemma 11.4.1 $G_0(t) \ge 0$ $(t \ge 0)$.

Now consider the function

$$P_2(\lambda) = K_0(\lambda) - me^{-\lambda}$$

with a constant m. It is the characteristic function of the equation

$$\left(\frac{d}{dt} + S_1\right)\left(\frac{d}{dt} + S_2\right)x(t) = mx(t-1) \ \ (t \ge 0). \tag{4.8}$$

By the integral inequalities (see Section 1.6), it is not hard to show that, if $G_0(t) \geq 0$ and $m \geq 0$, then the Green function of (4.8) is also non-negative.

Furthermore, assume that $P_2(\lambda) = K_2(\lambda)$, where $K_2(\lambda)$ is defined in Section 11.3. That is,

$$(\lambda+a_1+b_1e^{-\lambda})(\lambda+a_2+b_2e^{-\lambda})-me^{-\lambda} = \lambda^2+A\lambda+B\lambda e^{-\lambda}+C+De^{-\lambda}+Ee^{-2\lambda}.$$

Then, comparing the coefficients of $P_2(\lambda)$ and $K_2(\lambda)$, we get the relations

$$a_1 + a_2 = A, \quad a_1a_2 = C, \tag{4.9}$$

$$b_1 + b_2 = B, \quad b_1b_2 = E, \tag{4.10}$$

and

$$a_1b_2 + b_1a_2 - m = D. \tag{4.11}$$

Solving (4.9), we get

$$a_{1,2} = A/2 \pm (A^2/4 - C)^{1/2} = r_\pm(A, C).$$

Similarly, (4.10) implies

$$b_{1,2} = B/2 \pm (B^2/4 - E)^{1/2} = r_\pm(B, E).$$

From the hypothesis (3.7) it follows that $a_{1,2}, b_{1,2}$ are real. Condition (3.9) implies (4.7). So $G_0(t) \geq 0, t \geq 0$. Moreover, (3.8) provides relation (4.11) with a positive m.

But as it was mentioned above, if $G_0(t) \geq 0$, then the Green function $G_2(t)$ corresponding to $K_2(\lambda) = P_2(\lambda)$ is also positive. This proves the lemma. □

11.5 First-order nonlinear non-autonomous equations

First consider the linear equation

$$\dot{x}(t) + \int_0^\eta x(t-\tau)d_\tau\mu(t,\tau) = 0, \tag{5.1}$$

where $\mu(t,\tau)$ is a function defined on $[0,\infty) \times [0,\eta]$ nondecreasing in τ and continuous in t. Assume that there are nondecreasing functions $\mu_\pm(\tau)$ defined on $[0,\eta]$, such that

$$\mu_-(\tau_2) - \mu_-(\tau_1) \leq \mu(t,\tau_2) - \mu(t,\tau_1) \leq \mu_+(\tau_2) - \mu_+(\tau_1) \quad (\eta \geq \tau_2 > \tau_1 \geq 0). \tag{5.2}$$

We need also the autonomous equations

$$\dot{x}_+(t) + \int_0^\eta x_+(t-\tau)d\mu_+(\tau) = 0 \tag{5.3}$$

and

$$\dot{x}_-(t) + \int_0^\eta x_-(t-\tau)d\mu_-(\tau) = 0. \tag{5.4}$$

Denote by $G_1(t,s), G_+(t)$ and $G_-(t)$ the Green functions to (5.1), (5.3) and (5.4), respectively.

Lemma 11.5.1. *Let the conditions* (5.2) *and*

$$\mathrm{var}(\mu_+)\eta\, e < 1 \tag{5.5}$$

hold. Then

$$0 \le G_+(t-s) \le G_1(t,s) \le G_-(t-s) \quad (t \ge s \ge 0). \tag{5.6}$$

Proof. Due to Lemma 11.3.2, $G_\pm(t) \ge 0$ for all $t \ge 0$. In addition, according to the initial conditions,

$$G_+(0) = G_1(0,0) = G_-(0) = 1, \;\; G_+(t) = G_1(t,0) = G_-(t) = 0 \; (t < 0).$$

For a sufficiently small $t_0 > 0$, we have

$$G_1(t,0) \ge 0, \; \dot{G}_1(t,0) \le 0 \; (0 \le t \le t_0).$$

From (5.1) we obtain

$$\dot{G}_1(t,0) + \int_0^\eta G_1(t-\tau,0)d\mu_+(\tau) = f(t)$$

with

$$f(t) = \int_0^\eta G_1(t-\tau)(d\mu_+(\tau) - d_\tau\mu(t,\tau)).$$

Hence, by virtue of the Variation of Constants formula and (5.2), we arrive at the relation

$$G_1(t,0) = G_+(t) + \int_0^t G_+(t-s)f(s)ds \ge G_+(t) \; (0 \le t \le t_0).$$

Extending this inequality to the whole positive half-line, we get the left-hand part of inequality (5.6) for $s = 0$. Similarly the right-hand part of inequality (5.6) and the case $s > 0$ can be investigated. □

Note that due to (5.6)

$$\sup_t \int_0^t G_1(t,s)dt \le \sup_t \int_0^t G_-(t-s)ds = \int_0^\infty G_-(s)ds.$$

But by (5.4)

$$G_-(0) = 1 = \int_0^\infty \int_0^\eta G_-(t-\tau)d\mu_-(\tau)dt = \int_0^\eta \int_0^\infty G_-(t-\tau)dt\, d\mu_-(\tau)$$
$$= \int_0^\eta \int_{-\tau}^\infty G_-(s)dtd\mu_-(\tau) = \int_0^\eta \int_0^\infty G_-(s)dt\, d\mu_-(\tau)$$
$$= \int_0^\infty G_-(s)dt \ \operatorname{var}(\mu_-).$$

So we obtain the equality

$$\int_0^\infty G_\pm(s)ds = \frac{1}{\operatorname{var}(\mu_\pm)}. \tag{5.7}$$

Hence, we get

Corollary 11.5.2. *Let conditions* (5.2) *and* (5.5) *be fulfilled. Then*

$$G_1(t,s) \le 1 \ (t \ge s \ge 0),$$
$$0 \ge \frac{\partial G_1(t,s)}{\partial t} \ge -\int_0^\eta d_\tau \mu(t,\tau) = -\operatorname{var}\mu(t,.) \ \ (t \ge s),$$

and

$$\sup_t \int_0^t G_1(t,s)ds \le \frac{1}{\operatorname{var}(\mu_-)}. \tag{5.8}$$

Furthermore, consider the nonlinear equation

$$\dot x(t) + \int_0^\eta x(t-\tau)d_\tau\mu(t,\tau) = [Fx](t) \ \ (t>0), \tag{5.9}$$

where F is a causal mapping in $C(-\eta\,,\infty)$.

Denote

$$\Omega(r) := \{v \in C(-\eta\,,\infty) : \|v\|_{C(-\eta\,,\infty)} \le r\}$$

for a positive $r \le \infty$.

It is assumed that there is a constant q, such that

$$\|Fv\|_{C(-\eta\,,\infty)} \le q\, \|v\|_{C(-\eta\,,\infty)} \ (v \in \Omega(r)). \tag{5.10}$$

Following the arguments of the proof of Theorem 10.1.2, according to (5.8), we arrive at the main result of the present section.

Theorem 11.5.3. *Let the conditions* (5.2), (5.5), (5.10) *and*

$$q < \operatorname{var}(\mu_-) \tag{5.11}$$

hold. Then the zero solution of equation (5.9) *is stable.*

11.6 Comparison of Green's functions to second-order equations

Let us consider the equation

$$\ddot{u}(t) + 2c_0(t)\dot{u}(t) + c_1(t)\dot{u}(t-h) + d_0(t)u(t) + d_1(t)u(t-h) + d_2(t)u(t-2h) = 0 \quad (t \geq 0), \tag{6.1}$$

where $c_1(t), d_j(t)$ $(t \geq 0; j = 0, 1, 2)$ are piece-wise continuous functions, and $c_0(t)$ $(t \geq 0)$ is an absolutely continuous function having a piece-wise continuous derivative $\dot{c}_0(t)$.

Let $G(t,s)$ be the Green function to equation (6.1). So it is a function defined for $t \geq s - 2h$ $(s \geq 0)$, having continuous first and second derivatives in t for $t > s$, satisfying that equation for all $t > s \geq 0$ and the conditions

$$G(t,s) = 0 \;\; (s - 2h \leq t \leq s), \qquad \frac{\partial G(t,s)}{\partial t} = 0 \;\; (s - 2h \leq t < s);$$

and

$$\lim_{t \downarrow s} \frac{\partial G(t,s)}{\partial t} = 1.$$

Furthermore, extend $c_0(t)$ to $[-2h, \infty)$ by the relation

$$c_0(t) \equiv c_0(0) \text{ for } -2h \leq t \leq 0,$$

and put

$$a_1(t) = c_1(t)e^{\int_{t-h}^{t} c_0(s)ds}, \quad \text{and} \quad a_2(t) = d_2(t)e^{\int_{t-2h}^{t} c_0(s)ds} \;\; (t \geq 0).$$

The aim of this section is to prove the following result.

Theorem 11.6.1. *Let the conditions*

$$-\dot{c}_0(t) + c_0^2(t) + d_0(t) \leq 0 \tag{6.2}$$

and

$$-c_1(t)c_0(t-h) + d_1(t) \leq 0 \quad (t \geq 0) \tag{6.3}$$

hold. Let the Green function $G_0(t,s)$ to the equation

$$\ddot{u}(t) + a_1(t)\dot{u}(t-h) + a_2(t)u(t-2h) = 0 \;\; (t \geq 0) \tag{6.4}$$

be non-negative. Then the Green function $G(t,s)$ to equation (6.1) is also non-negative.

Proof. Substitute the equality

$$u(t) = w(t)e^{-\int_0^t c_0(s)ds} \tag{6.5}$$

into (6.1). Then, taking into account that

$$\frac{d}{dt}\int_0^{t-h} c_0(s)ds = \frac{d}{dt}\int_h^t c_0(s_1 - h)ds_1 = c_0(t-h),$$

we have

$$\begin{aligned} &e^{-\int_0^t c_0(s)ds}\Big[\ddot{w}(t) - 2c_0(t)\dot{w}(t) + w(t)(-\dot{c}_0(t) + c_0^2(t) + d_0(t)) \\ &\qquad + 2(c_0(t)\dot{w}(t) - c_0^2(t)w(t))\Big] \\ &+ c_1(t)e^{-\int_0^{t-h} c_0(s)ds}[-c_0(t-h)w(t-h) + \dot{w}(t-h)] \\ &+ d_1(t)e^{-\int_0^{t-h} c_0(s)ds}w(t-h) + d_2(t)e^{-\int_0^{t-2h} c_0(s)ds}w(t-2h) = 0. \end{aligned}$$

Or

$$\ddot{w}(t) + a_1(t)\dot{w}(t-h) + m_0(t)w(t) + m_1(t)w(t-h) + a_2(t)w(t-2h) = 0, \tag{6.6}$$

where

$$m_0(t) := -\dot{c}_0(t) + c_0^2(t) + d_0(t)$$

and

$$m_1(t) := e^{\int_{t-h}^t c_0(s)ds}[-c_1(t)c_0(t-h) + d_1(t)].$$

According to (6.3), $m_0(t) \le 0, m_1(t) \le 0$. Hence, by the integral inequalities principle, see Section 1.6, it easily follows that if the Green function to equation (6.4) is non-negative, then the Green function to equation (6.6) is also non-negative. This and (6.5) prove the theorem. □

11.7 Comments

This chapter is based on the papers [36, 49]. Theorem 11.6.1 is taken from the paper [48].

Recall that in 1949 M.A. Aizerman conjectured the following hypothesis: let A, b, c be an $n \times n$-matrix, a column-matrix and a row-matrix, respectively. Then for the absolute stability of the zero solution of the equation

$$\dot{x} = Ax + bf(cx) \quad (\dot{x} = dx/dt)$$

in the class of nonlinearities $f : \mathbb{R} \to \mathbb{R}$, satisfying the condition

$$0 \le f(s)/s \le q \quad (q = \text{const} > 0, s \in \mathbb{R},\ s \ne 0),$$

it is necessary and sufficient that the linear equation $\dot{x} = Ax + q_1 bcx$ be asymptotically stable for any $q_1 \in [0, q]$ [3]. These hypotheses caused great interest among specialists. Counterexamples were set up that demonstrated it was not, in general, true, cf. [107]. Therefore, the following problem arose: to find the class of systems that satisfy Aizerman's hypothesis. In 1983 the author had shown that any system satisfies the Aizerman hypothesis if its impulse function is non-negative. A similar result was proved for multivariable systems, distributed ones and in the input-output version. For the details see [24]. On the other hand, A.D. Myshkis [95, Section 10] pointed out the importance of consideration of the generalized Aizerman problem for retarded systems. In [27] that problem was investigated for retarded systems, whose nonlinearities have discrete constant delays; in [30], more general systems with nonlinearities acting in space C were considered.

The positivity conditions for fundamental solutions of first-order scalar differential equations are well known, cf. [1].

About recent very interesting results on absolute stability see, for instance, [86, 116] and references therein.

Chapter 12

Forced Oscillations in Vector Semi-Linear Equations

This chapter deals with forced periodic oscillations of coupled systems of semi-linear functional differential equations. Explicit conditions for the existence and uniqueness of periodic solutions are derived. These conditions are formulated in terms of the roots of characteristic matrix functions.

In addition, estimates for periodic solutions are established.

12.1 Introduction and statement of the main result

As as it is well known, any 1-periodic vector-valued function f with the property $f \in L^2(0,1)$ can be represented by the Fourier series

$$f = \sum_{k=-\infty}^{\infty} c_k e_k$$

where

$$e_k(t) = e^{2\pi i k t} \ (k = 0, \pm 1, \pm 2, \dots), \quad \text{and} \quad c_k = \int_0^1 f(t) e_k(t) dt \ \in \mathbb{C}^n$$

are the Fourier coefficients. Introduce the *Hilbert space PF of* 1*-periodic functions defined on the real axis* $\mathbb{R}$ *with values in* $\mathbb{C}^n$, *and the scalar product*

$$(f,u)_{PF} \equiv \sum_{k=-\infty}^{\infty} (c_k, b_k)_{C^n} \quad (f, u \in PF),$$

where $(.,.)_{C^n}$ is the scalar product $\mathbb{C}^n$, c_k and b_k are the Fourier coefficients of f and u, respectively. The norm in PF is

$$|f|_{PF} = \sqrt{(f,f)_{PF}} = \left(\sum_{k=-\infty}^{\infty} \|c_k\|_n^2 \right)^{1/2}.$$

Here $\|c\|_n = \sqrt{(c,c)_{C^n}}$ for a vector c.

Due to the periodicity, for any real a we have,

$$\|v\|_{L^2([0,1],\mathbb{C}^n)} = \left[\int_a^{a+1} \|v(t)\|_n^2 dt\right]^{1/2} \quad (v \in PF).$$

The Parseval equality yields

$$|v|_{PF} = \|v\|_{L^2([0,1],\mathbb{C}^n)} = \left[\int_a^{a+1} \|v(t)\|_n^2 dt\right]^{1/2} \quad \text{for any real } a.$$

Let $R(\tau)$ be an $n \times n$-matrix-valued function of bounded variation defined on $[0,1]$. Consider the system

$$\dot{u}(t) - \int_0^1 dR(\tau)u(t-\tau) + (Fu)(t) \quad (t \in \mathbb{R}) \tag{1.1}$$

where $F : PF \to PF$ is a mapping satisfying the condition

$$|Fv - Fw|_{PF} \le q|v - w|_{PF} \ (q = \text{const} > 0;\ v, w \in PF). \tag{1.2}$$

In addition,

$$l := |F0|_{PF} > 0. \tag{1.3}$$

That is, $F0$ is not zero identically.

In particular, one can take $Fv = \hat{F}v + f$, where $\hat{F}$ is causal and $f \in PF$.

Let $K(z)$ be the characteristic matrix of the linear term of (1.1):

$$K(z) = zI - \int_0^1 e^{-z\tau} dR(\tau) \quad (z \in \mathbb{C}),$$

where I is the unit matrix. Assume that matrices $K(2i\pi j)$ are invertible for all integer j and the condition

$$M_0(K) := \sup_{j=0,\pm1,\pm2,\dots} \|K^{-1}(2i\pi j)\|_n < \infty$$

hold. Here the matrix norm is spectral.

An absolutely continuous function $u \in PF$ satisfying equation (1.1) *for almost all $t \in \mathbb{R}$ will be called a periodic solution to that equation.* A solution is nontrivial if it is not equal to zero identically. Now we are in a position to formulate the main result of the chapter.

Theorem 12.1.1. *Let the conditions* (1.2), (1.3) *and*

$$qM_0(K) < 1 \tag{1.4}$$

be fulfilled. Then equation (1.1) *has a unique nontrivial periodic solution u. Moreover, it satisfies the estimate*

$$|u|_{PF} \le \frac{lM_0(K)}{1 - qM_0(K)}. \tag{1.5}$$

The proof of this theorem is presented in the next section.

Note that

$$M_0(K) \le \sup_{\omega \in \mathbb{R}} \|K^{-1}(i\omega)\|_n.$$

Now Lemma 4.3.1 implies

$$M_0(K) \le \theta(K) := \sup_{|\omega| \le 2\,\mathrm{var}(R)} \|K^{-1}(i\omega)\|_n.$$

So one can use the estimates for $\theta(K)$ derived in Sections 4.4 and 4.5.

12.2 Proof of Theorem 12.1.1

In this section for brevity we set $M_0(K) = M_0$. Furthermore, consider an operator T defined on PF by the equality

$$(Tw)(t) = \dot{w}(t) - \int_0^1 dR(\tau)w(t-\tau) \; (t \in \mathbb{R},\ w \in W(PF)).$$

We need the linear equation

$$(Tw)(t) = f(t) \tag{2.1}$$

with $f \in PF$, $t \in \mathbb{R}$. For any $h \in \mathbb{C}^n$ and an integer k we have

$$(T(he_k))(t) = e_k(t)\left[2i\pi kI - \int_0^1 e^{-2i\pi k\tau} dR(\tau)\right] h = e_k(t)K(2\pi ik)h.$$

We seek a solution u to (2.1) in the form

$$u = \sum_{k=-\infty}^{\infty} a_k e_k \quad (a_k \in \mathbb{C}^n).$$

Hence,

$$Tu = \sum_{k=-\infty}^{\infty} Te_k a_k e_k = \sum_{k=-\infty}^{\infty} e_k K(2\pi ik)a_k$$

and by (2.1),

$$a_k = K^{-1}(2\pi ik)c_k \quad (k = 0, \pm1, \pm2, \dots).$$

Therefore,

$$\|a_k\|_n \le M_0\|c_k\|_n.$$

Hence, $|u|_{PF} \le M_0|f|_{PF}$. This means that $|T^{-1}|_{PF} \le M_0$.

Furthermore, equation (1.1) is equivalent to

$$u = \Psi(u) \tag{2.2}$$

where

$$\Psi(u) = T^{-1}Fu.$$

For any $v, w \in PF$ relation (1.2) implies

$$|\Psi(v) - \Psi(w)|_{PF} \le M_0 q |v - w|_{PF}.$$

Due to (1.3) Ψ maps PF into itself. So according to (1.4) and the Contraction Mapping theorem equation (1.1) has a unique solution $u \in PF$. To prove estimate (1.5), note that

$$|u|_{PF} = |\Psi(u)|_{PF} \le M_0(q|u|_{PF} + l).$$

Hence (1.4) implies the required result. □

12.3 Applications of matrix functions

Consider the problem

$$y'(t) = Ay(t) + [Fy](t), \ y(0) = y(1), \tag{3.1}$$

where A is a constant invertible diagonalizable matrix (see Section 2.7) and F is an arbitrary continuous mapping of PF into itself, satisfying

$$|Fy|_{PF} \le q|y|_{PF} + l \ \ (q, l = \text{const} > 0) \quad \text{and} \quad F(0) \ne 0. \tag{3.2}$$

For example, let $(Fy)(t) = B(t)F_0(y(t-1))$, where $B(t)$ is a 1-periodic matrix-function and $F_0 : \mathbb{R}^n \to \mathbb{R}^n$ has the Lipschitz property and $F_0(0) \ne 0$.

First, consider the scalar problem

$$x'(t) = \omega x(t) + f_0(t) \quad (x(0) = x(1)) \tag{3.3}$$

with a constant $\omega \ne 0$ and a function $f_0 \in L^2([0,1], \mathbb{C})$. Then a solution of (3.3) is given by

$$x(t) = \int_0^1 G(\omega, t, s) f_0(s) ds, \tag{3.4}$$

where

$$G(\omega, t, s) = \frac{1}{1 - e^{\omega}} \begin{cases} e^{\omega(1+t-s)} & \text{if } 0 \le t \le s \le 1, \\ e^{\omega(t-s)} & \text{if } 0 \le s < t \le 1 \end{cases}$$

is the Green function to the problem (3.3), cf. [70].

Now consider the vector equation

$$w'(t) = Aw(t) + f(t), \ y(0) = y(1)$$

with an $f \in PF$. Then a solution of this equation is given by

$$w(t) = \int_0^1 G(A, t, s) f(s) ds. \tag{3.5}$$

Since A is a diagonalizable matrix, we have

$$G(A,t,s) = \sum_{k=1}^{n} G(\lambda_k(A),t,s)P_k$$

where P_k are the Riesz projections and $\lambda_k(A)$ are the eigenvalues of A (see Section 2.7).

By Corollary 2.7.5,

$$\|G(A,t,s)\|_n \le \gamma(A) \max_k |G(\lambda_k(A),t,s)|.$$

Consequently,

$$\int_0^1 \int_0^1 \|G(A,t,s)\|_n^2 ds dt \le J_A^2,$$

where

$$J_A = \gamma(A) \max_k \left[\int_0^1 \int_0^1 |G(\lambda_k(A),t,s)|^2 ds \; dt \right]^{1/2}.$$

So the operator $\hat{G}$ defined by

$$\hat{G}f(t) = \int_0^1 G(A,t,s)f(s)ds$$

is a Hilbert–Schmidt operator with the Hilbert–Schmidt norm $N_2(\hat{G}) \le J_A$. By the Parseval equality we can write $|\hat{G}|_{PF} \le J_A$.

Rewrite (3.1) as

$$y(t) = \int_0^1 G(A,t,s)[Fy](s)ds.$$

Hence condition (3.2) implies the inequality

$$|y|_{PF} \le J_A(q|y|_{PF} + l).$$

Due to the Schauder fixed point theorem, we thus arrive at the following result.

Theorem 12.3.1. *Let the conditions* (3.2) *and*

$$J_A q < 1$$

hold. Then (3.1) *has a nontrivial solution* y. *Moreover, it satisfies the inequality*

$$|y|_{PF} \le \frac{J_A l}{1 - J_A q}.$$

12.4 Comments

The material of this chapter is adapted from the paper [28].

Periodic solutions (forced periodic oscillations) of nonlinear functional differential equations (FDEs) have been studied by many authors, see for instance [10, 66, 79] and references given therein. In many cases, the problem of the existence of periodic solutions of FDEs is reduced to the solvability of the corresponding operator equations. But for the solvability conditions of the operator equations, estimates for the Green functions of linear terms of equations are often required. In the general case, such estimates are unknown. Because of this, the existence results were established mainly for semilinear coupled systems of FDEs.

Chapter 13

Steady States of Differential Delay Equations

In this chapter we investigate steady states of differential delay equations. Steady states of many differential delay equations are described by equations of the type

$$F_0(x) = 0,$$

where $F_0 : \mathbb{C}^n \to \mathbb{C}^n$ is a function satisfying various conditions. For example consider the equation

$$\dot{y}(t) = F_1(y(t), y(t-h)),$$

where F_1 maps $\mathbb{C}^n \times \mathbb{C}^n$ into $\mathbb{C}^n$. Then the condition $y(t) \equiv x \in \mathbb{C}^n$ yields the equation $F_1(x, x) = 0$. So in this case $F_0(x) = F_1(x, x)$.

13.1 Systems of semilinear equations

In this chapter $\|.\|$ is the Euclidean norm and

$$\Omega(r; \mathbb{C}^n) := \{x \in \mathbb{C}^n : \|x\| \le r\}$$

for a positive $r \le \infty$.

Let us consider in $\mathbb{C}^n$ the nonlinear equation

$$Ax = F(x), \tag{1.1}$$

where A is an invertible matrix, and F continuously maps $\Omega(r; \mathbb{C}^n)$ into $\mathbb{C}^n$.

Assume that there are positive constants q and l, such that

$$\|F(h)\| \le q\|h\| + l \quad (h \in \Omega(r; \mathbb{C}^n)). \tag{1.2}$$

Lemma 13.1.1. *Under condition* (1.2) *with* $r < \infty$, *let*

$$\|A^{-1}\|(qr + l) \leq r. \tag{1.3}$$

Then equation (1.1) *has at least one solution* $x \in \Omega(r; \mathbb{C}^n)$, *satisfying the inequality*

$$\|x\| \leq \frac{\|A^{-1}\| l}{1 - q\|A^{-1}\|}. \tag{1.4}$$

Proof. Set

$$\Psi(y) = A^{-1}F(y) \ \ (y \in \mathbb{C}^n).$$

Hence,

$$\|\Psi(y)\| \leq \|A^{-1}\|(q\|y\| + l) \leq \|A^{-1}\|(qr + l) \leq r \ (y \in \Omega(r; \mathbb{C}^n)). \tag{1.5}$$

So due to the Brouwer Fixed Point theorem, equation (1.1) has a solution. Moreover, due to (1.3),

$$\|A^{-1}\|q < 1.$$

Now, using (1.5), we easily get (1.4). □

Put

$$R(A) = \sum_{k=0}^{n-1} \frac{g^k(A)}{d_0^{k+1}(A)\sqrt{k!}}$$

where $g(A)$ is defined in Section 2.3, $d_0(A)$ is the lower spectral radius. That is, $d_0(A)$ is the minimum of the absolute values of the eigenvalues of A:

$$d_0(A) := \min_{k=1,\dots,n} |\lambda_k(A)|.$$

Due to Corollary 2.3.3,

$$\|A^{-1}\| \leq R(A).$$

Now the previous lemma implies

Theorem 13.1.2. *Under condition* (1.2), *let*

$$R(A)(qr + l) \leq r.$$

Then equation (1.1) *has at least one solution* $x \in \Omega(r; \mathbb{C}^n)$, *satisfying the inequality*

$$\|x\| \leq \frac{R(A)\, l}{1 - qR(A)}.$$

13.2 Essentially nonlinear systems

Consider the coupled system

$$\sum_{k=1}^{n} a_{jk}(x)x_k = f_j \quad (j = 1, \dots, n;\ x = (x_j)_{j=1}^n \in \mathbb{C}^n\ ), \tag{2.1}$$

where

$$a_{jk} : \Omega(r; \mathbb{C}^n) \to \mathbb{C} \ \ (j, k = 1, \dots, n)$$

are continuous functions and $f = (f_j) \in \mathbb{C}^n$ is given. We can write system (2.1) in the form

$$A(x)x = f \tag{2.2}$$

with the matrix

$$A(z) = (a_{jk}(z))_{j,k=1}^n \ \ (z \in \Omega(r; \mathbb{C}^n)).$$

Theorem 13.2.1. *Let*

$$\inf_{z\in\Omega(r;\mathbb{C}^n)} d_0(A(z)) \equiv \inf_{z\in\Omega(r;\mathbb{C}^n)} \min_k |\lambda_k(A(z))| > 0$$

and

$$\theta_r := \sup_{z\in\Omega(r;\mathbb{C}^n)} \sum_{k=0}^{n-1} \frac{g^k(A(z))}{\sqrt{k!}\,d_0^{k+1}(A(z))} \le \frac{r}{\|f\|}. \tag{2.3}$$

Then system (2.1) *has at least one solution* $x \in \Omega(r; \mathbb{C}^n)$, *satisfying the estimate*

$$\|x\| \le \theta_r \|f\|. \tag{2.4}$$

Proof. Thanks to Corollary 2.3.3,

$$\|A^{-1}(z)\| \le \theta_r \ \ (z \in \Omega(r; \mathbb{C}^n)).$$

Rewrite (2.2) as

$$x = \Psi(x) \equiv A^{-1}(x)f. \tag{2.5}$$

Due to (2.3)

$$\|\Psi(z)\| \le \theta_r \|f\| \le r \ \ (z \in \Omega(r; \mathbb{C}^n)).$$

So Ψ maps $\Omega(r; \mathbb{C}^n)$ into itself. Now the required result is due to the Brouwer Fixed Point theorem. □

Corollary 13.2.2. *Let matrix* $A(z)$ *be normal:*

$$A^*(z)A(z) = A(z)A^*(z) \ \ (z \in \Omega(r; \mathbb{C}^n)).$$

If, in addition,

$$\|f\| \le r \inf_{z\in\Omega(r;\mathbb{C}^n)} d_0(A(z)), \tag{2.6}$$

then system (2.1) *has at least one solution* x *satisfying the estimate*

$$\|x\| \le \frac{\|f\|}{\inf_{z\in\Omega(r;\mathbb{C}^n)} d_0(A(z))}.$$

Indeed, if $A(z)$ is normal, then $g(A(z)) \equiv 0$ and

$$\theta_r = \frac{1}{\inf_{z \in \Omega(r;\mathbb{C}^n)} d_0(A(z))}.$$

Corollary 13.2.3. *Let matrix $A(z)$ be upper triangular:*

$$\sum_{k=j}^{n} a_{jk}(x)x_k = f_j \quad (j = 1, \dots, n). \tag{2.7}$$

In addition to the notation

$$\tau(A(z)) := \sum_{1 \le j < k \le n} |a_{jk}(z)|^2, \quad \textit{and} \quad \tilde{a}(z) := \min_{j=1,\dots,n} |a_{jj}(A(z)|,$$

let

$$\sup_{z \in \Omega(r;\mathbb{C}^n)} \sum_{k=0}^{n-1} \frac{\tau^k(A(z))}{\sqrt{k!}\tilde{a}^{k+1}(z)} \le \frac{r}{\|f\|}. \tag{2.8}$$

Then system (2.7) *has at least one solution $x \in \Omega(r; \mathbb{C}^n)$.*

Indeed, this result is due to Theorem 13.2.1, since the eigenvalues of a triangular matrix are its diagonal entries, and

$$g(A(z)) \le \tau(A(z)) \quad \text{and} \quad d_0(A(z)) = \tilde{a}(z).$$

A similar result is true for lower triangular systems.

Note that according to the relation

$$g(A_0) \le \sqrt{1/2} N_2(A_0^* - A_0)$$

for any constant matrix A_0 (see Section 2.3), in the general case, $g(A(z))$ can be replaced by the simply calculated quantity

$$v(A(z)) = \sqrt{1/2} N_2(A^*(z) - A(z)) = \left[\frac{1}{2} \sum_{k=1}^{n} |a_{jk}(z) - \overline{a}_{kj}(z)|^2 \right]^{1/2}. \tag{2.9}$$

Example 13.2.4. Let us consider the system

$$a_{j1}(x)x_1 + a_{j2}(x)x_2 = f_j \quad (x = (x_1, x_2) \in \mathbb{C}^2), \tag{2.10}$$

where f_j are given numbers, and continuous scalar-valued functions

$$a_{jk} \quad (j, k = 1, 2)$$

are defined on

$$\Omega(r; \mathbb{C}^2) \equiv \{z \in \mathbb{C}^2 : \|z\| \le r\}.$$

Due to (2.9)

$$g(A(z)) \le |a_{21}(z) - \overline{a}_{12}(z)|.$$

In addition, $\lambda_{1,2}(A(z))$ are the roots of the polynomial

$$y^2 - t(z)y + b(z),$$

where

$$t(z) = \text{Trace}(A(z)) = a_{11}(z) + a_{22}(z)$$

and

$$b(z) = \det\,(A(z)) = a_{11}(z)a_{22}(z) - a_{12}(z)a_{21}(z).$$

Then

$$d_0(A(z)) := \min_{k=1,2} |\lambda_k(A(z))|.$$

Moreover, if $\inf_{z\in\Omega(r;\mathbb{C}^2)} d_0(A(z)) > 0$, then

$$\theta_r \le \tilde{\theta}_r := \sup_{z\in\Omega(r;\mathbb{C}^2)} \frac{1}{d_0(A(z))} + \frac{|a_{21}(z) - \overline{a}_{12}(z)|}{d_0^2(A(z))}.$$

If

$$\|f\|\tilde{\theta}_r \le r,$$

then due to Theorem 13.2.1, system (2.10) has a solution.

13.3 Nontrivial steady states

Consider the system

$$F_j(x) = 0 \quad (j = 1, \ldots, n;\ x = (x_j)_{j=1}^n \in \mathbb{C}^n\ ), \tag{3.1}$$

where functions $F_j : \Omega(r; \mathbb{C}^n) \to \mathbb{C}$ admit the representation

$$F_j(x) = \psi_j(x) \left(\sum_{k=1}^n a_{jk}(x)x_k - f_j \right) \quad (j = 1, \ldots, n). \tag{3.2}$$

Here $\psi_j : \Omega(r; \mathbb{C}^n) \to \mathbb{C}$ are functions with the property $\psi_j(0) = 0$, $a_{jk} : \Omega(r; \mathbb{C}^n) \to \mathbb{C}$ are continuous functions, and f_j $(j, k = 1, \ldots, n)$ are given numbers. Again put

$$A(z) = (a_{jk}(z))_{j,k=1}^n \quad (z \in \Omega(r; \mathbb{C}^n))$$

assuming that it is invertible on $\Omega(r; \mathbb{C}^n)$.

In the rest of this section *it is assumed that at least one of the numbers f_j is non-zero.* Then

$$\|A(\hat{x})\|\|\hat{x}\| \ge \|f\| > 0$$

for any solution $\hat{x} \in \Omega(r; \mathbb{C}^n)$ of equation (2.2) (if it exists). So $\hat{x}$ is non-trivial. Obviously, (3.1) has the trivial (zero) solution. Moreover, if (2.2) has a solution $\hat{x} \in \Omega(r; \mathbb{C}^n)$, then it is simultaneously a solution of (3.1).

Now Theorem 13.2.1 implies

Theorem 13.3.1. *Let condition* (2.3) *hold. Then system* (3.1) *with* F_j *defined by* (3.2) *has at least two solutions: the trivial solution and a nontrivial one satisfying inequality* (2.4).

In addition, the previous theorem and Corollary 13.2.2 yield

Corollary 13.3.2. *Let matrix* $A(z)$ *be normal for any* $z \in \Omega(r; \mathbb{C}^n)$ *and condition* (2.6) *hold. Then system* (3.1) *has at least two solutions: the trivial solution and a nontrivial one belonging to* $\Omega(r; \mathbb{C}^n)$.

Theorem 13.3.2 and Corollary 13.2.3 imply

Corollary 13.3.3. *Let matrix* $A(z)$ *be upper triangular for any* $z \in \Omega(r; \mathbb{C}^n)$. *Then under condition* (2.8), *system* (3.1) *has at least two solutions: the trivial solution and a nontrivial one belonging to* $\Omega(r; \mathbb{C}^n)$.

A similar result is valid if $A(z)$ is lower triangular.

Example 13.3.4. Let us consider the system

$$\psi_j(x_1, x_2)(a_{j1}(x)x_1 + a_{j2}(x)x_2 - f_j) = 0 \quad (x = (x_1, x_2) \in \mathbb{C}^2) \tag{3.3}$$

where functions $\psi_j : \Omega(r; \mathbb{C}^n) \to \mathbb{C}$ have the property $\psi_j(0,0) = 0$. In addition, f_j, a_{jk} $(j, k = 1, 2)$ are the same as in Example 13.2.4. Assume that at least one of the numbers f_1, f_2 is non-zero. Then due to Theorem 13.3.1, under conditions (2.3), system (3.3) has in $\Omega(r; \mathbb{C}^n)$ at least two solutions.

13.4 Positive steady states

Consider the coupled system

$$u_j - \sum_{k=1,\ k \neq j}^{n} a_{jk}(u)u_k = F_j(u) \quad (j = 1, \dots, n), \tag{4.1}$$

where

$$a_{jk}, F_j : \Omega(r; \mathbb{C}^n) \to \mathbb{R} \quad (j \neq k;\ j, k = 1, \dots, n)$$

are continuous functions. For instance, the coupled system

$$\sum_{k=1}^{n} w_{jk}(u)u_k = f_j \quad (j = 1, \dots, n), \tag{4.2}$$

where f_j are given real numbers, $w_{jk} : \Omega(r;\mathbb{C}^n) \to \mathbb{R}$ are continuous functions, can be reduced to (4.1) with

$$a_{jk}(u) \equiv -\frac{w_{jk}(u)}{w_{jj}(u)} \quad \text{and} \quad F_j(u) \equiv \frac{f_j}{w_{jj}(u)},$$

provided

$$w_{jj}(z) \neq 0 \ (z \in \Omega(r;\mathbb{C}^n);\ j = 1, \dots, n). \tag{4.3}$$

Put

$$c_r(F) = \sup_{z \in \Omega(r;\mathbb{C}^n)} \|F(z)\|.$$

Denote

$$V_+(z) := \begin{pmatrix} 0 & a_{12}(z) & \dots & a_{1n}(z) \\ 0 & 0 & \dots & a_{2n}(z) \\ . & \dots & . & . \\ 0 & 0 & \dots & 0 \end{pmatrix},$$

$$V_-(z) := \begin{pmatrix} 0 & \dots & 0 & 0 \\ a_{21}(z) & \dots & 0 & 0 \\ . & \dots & . & \\ a_{n1}(z) & \dots & a_{n,n-1}(z) & 0 \end{pmatrix}.$$

Recall that $N_2(A)$ is the Frobenius norm of a matrix A. So

$$N_2^2(V_+(z)) = \sum_{j=1}^{n-1} \sum_{k=j+1}^{n} a_{jk}^2(z), \quad N_2^2(V_-(z)) = \sum_{j=1}^{n} \sum_{k=2}^{j-1} a_{jk}^2(z).$$

In addition, put

$$\tilde{J}_{R^n}(V_\pm(z)) \equiv \sum_{k=0}^{n-1} \frac{N_2^k(V_\pm(z))}{\sqrt{k!}}.$$

Theorem 13.4.1. *Let the conditions*

$$\alpha_r \equiv \max\{\inf_{z\in\Omega_r} (\frac{1}{\tilde{J}_{R^n}(V_-(z))} - \|V_+(z)\|),\ \inf_{z\in\Omega_r} (\frac{1}{\tilde{J}_{R^n}(V_+(z))} - \|V_-(z)\|)\} > 0$$

and

$$c_r(F) < r\alpha_r(\Omega_r)$$

hold. Then system (4.1) *has at least one solution* $u \in \Omega(r;\mathbb{C}^n)$ *satisfying the inequality*

$$\|u\| \leq \frac{c_r(F)}{\alpha_r(\Omega_r)}.$$

In addition, let $a_{jk}(z) \geq 0$ *and* $F_j(z) \geq 0$ $(j \neq k;\ j,k = 1, \dots, n)$ *for all* z *from the ball*

$$\{z \in \mathbb{R}^n : \|z\| \leq \frac{c_r(F)}{\alpha_r(\Omega_r)}\}.$$

Then the solution u *of* (4.1) *is non-negative*

For the proof see [34].

13.5 Systems with differentiable entries

Consider the system

$$f_k(y_1,\dots,y_n)=h_k\in\mathbb{C}\ (k=1,\dots,n),$$

where $f_j(x)=f_j(x_1,x_2,\dots,x_n)$, $f_j(0)=0\quad(j=1,\dots,n)$ are scalar-valued continuously differentiable functions defined on $\mathbb{C}^n$. Put $F(x)=(f_j(x))_{j=1}^n$ and

$$F'(x)=\left(\frac{\partial f_i(x)}{\partial x_j}\right)_{i,j=1}^n.$$

That is, $F'(x)$ is the Jacobian matrix. Rewrite the considered system as

$$F(y)=h\in\mathbb{C}^n. \tag{5.1}$$

For a positive number $r\le\infty$ assume that

$$\rho_0(r)\equiv\min_{x\in\Omega(r;\mathbb{C}^n)}d_0(F'(x))=\min_{x\in\Omega(r;\mathbb{C}^n)}\min_k|\lambda_k(F'(x))|>0, \tag{5.2}$$

and

$$\tilde g_0(r)=\max_{x\in\Omega(r;\mathbb{C}^n)}g(F'(x))<\infty. \tag{5.3}$$

Finally put

$$p(F,r)\equiv\sum_{k=0}^{n-1}\frac{\tilde g_0^k(r)}{\sqrt{k!}\rho_0^{k+1}(r)}.$$

Theorem 13.5.1. *Let $f_j(x)=f_j(x_1,x_2,\dots,x_n)$, $f_j(0)=0\quad(j=1,\dots,n)$ be scalar-valued continuously differentiable functions, defined on $\Omega(r;\mathbb{C}^n)$. Assume that conditions* (5.2) *and* (5.3) *hold. Then for any $h\in\mathbb{C}^n$ with the property*

$$\|h\|\le\frac{r}{p(F,r)},$$

there is a solution $y\in\mathbb{C}^n$ of system (5.1) *which subordinates to the inequality*

$$\|y\|\le\frac{\|h\|}{p(F,r)}.$$

For the proof of this result see [22].

13.6 Comments

This chapter is based on the papers [22] and [34].

Chapter 14

Multiplicative Representations of Solutions

In this chapter we suggest a representation for solutions of differential delay equations via multiplicative operator integrals.

14.1 Preliminary results

Let X be a Banach space with the unit operator I, and A be a bounded linear operator acting in X.

A family $P(t)$ of projections in X defined on a finite real segment $[a,b]$ is called *a resolution of the identity,* if it satisfies the following conditions:

1) $P(a) = 0, P(b) = I$,

2) $P(t)P(s) = P(\min\{t,s\})$ $(t, s \in [a,b])$

and

3) $\sup_{t\in[a,b]} \|P(t)\|_X < \infty$.

A resolution of the identity $P(t)$ is said to be the *spectral resolution* of A, if

$$P(t)AP(t) = AP(t) \quad (t \in [a,b]). \tag{1.1}$$

We will say that A *has a vanishing diagonal,* if it has a spectral resolution $P(t)$ $(a \le t \le b)$, and with the notation

$$\Delta P_k = \Delta P_{k,n} = P(t_k^{(n)}) - P(t_{k-1}^{(n)}) \ (k = 1, \ldots, n;\ a = t_0^{(n)} < t_1^{(n)} < \cdots < t_n^{(n)} = b),$$

the sums

$$D_n := \sum_{k=1}^{n} \Delta P_k A \Delta P_k$$

tend to zero in the operator norm, as $\max_k |t_k^{(n)} - t_{k-1}^{(n)}| \to 0$.

Lemma 14.1.1. *Let a bounded linear operator A acting in X have a spectral resolution $P(t)$ $(a \le t \le b)$ and a vanishing diagonal. Then the sequence of the operators*

$$Z_n = \sum_{k=1}^{n} P(t_{k-1}) A \Delta P_k$$

converges to A in the operator norm as $\max_k |t_k^{(n)} - t_{k-1}^{(n)}|$ tends to zero.

Proof. Thanks to (1.1) $\Delta P_j A \Delta P_k$ for $j > k$. So

$$A = \sum_{j=1}^{n} \sum_{k=1}^{n} \Delta P_j A \Delta P_k = \sum_{k=1}^{n} \sum_{j=1}^{k} \Delta P_j A \Delta P_k = Z_n + D_n.$$

By the assumption $D_n \to 0$ as $n \to \infty$. Thus $Z_n - A = D_n \to 0$ as $n \to \infty$, as claimed. □

For bounded operators $A_1, A_2, \ldots, A_j$ put

$$\prod_{1 \le k \le j}^{\rightarrow} A_k \equiv A_1 A_2, \ldots A_j.$$

That is, the arrow over the symbol of the product means that the indexes of the co-factors increase from left to right.

Lemma 14.1.2. *Let $\{\tilde{P}_k\}_{k=0}^{n}$ $(n < \infty)$ be an increasing chain of projections in X. That is,*

$$0 = \text{range}\, \tilde{P}_0 \subset \text{range}\, \tilde{P}_{n-1} \subset \cdots \subset \text{range}\ \tilde{P}_n = I.$$

Suppose a linear operator W in X satisfies the relation

$$\tilde{P}_{k-1} W \tilde{P}_k = W \tilde{P}_k \text{ for } k = 1, 2, \ldots, n.$$

Then W is a nilpotent operator and

$$(I - W)^{-1} = \prod_{2 \le k \le n}^{\rightarrow} (I + W_k)$$

with $W_k = W(\tilde{P}_k - \tilde{P}_{k-1})$.

Proof. Taking into account that $\tilde{P}_n = I$ and $W\tilde{P}_1 = 0$, we can write down

$$W^j = W \tilde{P}_{n-j+1} W \tilde{P}_{n-j+2} \cdots \tilde{P}_{n-1} W \tilde{P}_n.$$

Hence

$$W^n = 0, \tag{1.2}$$

i.e., W is a nilpotent operator. Besides,

$$W\tilde{P}_m = \sum_{k=1}^{m} W_k \ (m \le n)$$

and thus

$$W^j = \sum_{2\le k_1<k_2<\cdots<k_j\le n} W_{k_1}W_{k_2}\dots W_{k_j}.$$

Furthermore, we have

$$\prod_{2\le k\le n}^{\rightarrow} (I+W_k) = I + \sum_{k=2}^{n} W_k + \sum_{2\le k_1<k_2\le n} W_{k_1}W_{k_2} + \cdots + W_2W_3\dots W_n.$$

Simple calculations show that

$$\prod_{2\le k\le n}^{\rightarrow} (I+W_k) = \sum_{j=1}^{n-1} W^j.$$

But thanks to (1.2) it follows that

$$(I-W)^{-1} = \sum_{j=1}^{n-1} W^j.$$

This proves the result. □

Furthermore, let $P(t)$ be a resolution of the identity in X and let $F(t)$ be a function defined on $[a,b]$, whose values are bounded operators in X and continuous in t in the operator topology. Introduce *the right multiplicative integral*

$$\int_{[a,b]}^{\rightarrow} (I + F(s)\,dP(s))$$

as the limit in the operator norm (if it exists) of the sequence of the products

$$\prod_{1\le k\le n}^{\rightarrow} (I + F(t_k)\Delta P_k),$$

as $\max_k |t_k^{(n)} - t_{k-1}^{(n)}|$ tends to zero. In particular,

$$\int_{[a,b]}^{\rightarrow} (I + A\,dP(s))$$

denotes the limit in the operator norm of the sequence of the products

$$\prod_{1\le k\le n}^{\rightarrow} (I + A\Delta P_k).$$

Similarly, *the left multiplicative integral* is defined as the limit in the operator norm (if it exists) of the sequence of the products

$$\prod_{1\le k\le n}^{\leftarrow} (I+F(t_k)\Delta P_k) = (I+F(t_n)\Delta P_n)(I+F(t_{n-1})\Delta P_{n-1})\dots(I+F(t_1)\Delta P_1).$$

The left multiplicative integral is denoted by

$$\int_{[a,b]}^{\leftarrow} (I+F(s)\,dP(s)).$$

Lemma 14.1.3. *Let a linear operator A have a continuous spectral resolution $P(t)$ defined on a segment $[a,b]$ and a vanishing diagonal. Then*

$$(I-A)^{-1} = \int_{[a,b]}^{\rightarrow} (1+AdP(t)).$$

Proof. By Lemma 14.1.1, A is the limit in the operator norm of Z_n. Due to Lemma 14.1.2,

$$(I-Z_n)^{-1} = \prod_{1\le k\le n}^{\rightarrow} (I+Z_n\Delta P_k).$$

Hence letting $n\to\infty$, we get the required result. □

14.2 Volterra equations

Let $\mathbb{C}^n$ be a complex Euclidean space and $C(0,T) = C([0,T],\mathbb{C}^n)$ $(T<\infty)$. Consider in $\mathbb{C}^n$ the equation

$$x(t) - \int_0^t K(t,s)x(s)ds = f(t) \tag{2.1}$$

where $K(t,s)$ is a matrix kernel, such that

$$K(t,s) = 0 \ (0\le s<t\le T). \tag{2.2}$$

In addition, $K(t,s)$ is continuous in t and Reimann-integrable in s, and f is Reimann-integrable. It is not hard to check the operator V defined by

$$(Vw)(t) = \int_0^t K(t,s)w(s)\,ds \quad (w\in C(0,T)) \tag{2.3}$$

is a compact operator in $C(0,T)$.

For each $\tau\in(0,T)$ we define the projection $Q(\tau)$ by

$$(Q(\tau)w)(t) := \begin{cases} 0 & \text{if} \quad \tau<t\le T, \\ w(t) & \text{if} \quad 0\le t\le\tau. \end{cases}$$

In addition, $Q(0) = 0$ and $Q(T) = I$ – the identity operator in $C(0,T)$. Then,

$$Q(\tau)VQ(\tau) = Q(\tau)V, \quad \tau \in [0,T],$$

$V\,dP(\tau)w(t) = 0$ if $t \neq \tau$ and

$$V\,dP(\tau)w(\tau) = K(t,\tau)w(\tau)d\tau \ (w \in C(0,T)).$$

If we put $\hat{P}(\tau) = I - Q(T-\tau)$, then we obtain

$$(I - \hat{P}(\tau))V(I - \hat{P}(\tau)) = (I - \hat{P}(\tau))V.$$

Otherwise $\hat{P}(\tau)V\hat{P}(\tau) = \hat{V}P(\tau)$. Consequently, $\hat{P}(t)$ is the spectral resolution of V. Moreover, relation (2.2) and simple calculations show that V has a vanishing diagonal with respect to $\hat{P}(\tau)$.

So we can use Lemma 14.1.3. It asserts that $(I-V)^{-1}$ is the limit in the operator norm of the products

$$\prod_{2\leq k\leq m}^{\rightarrow} (I + V\Delta\hat{P}(t_k))$$

which are equivalent to the products

$$\prod_{1\leq k\leq m-1}^{\leftarrow} (I + V\Delta Q(t_k)).$$

So we have proved the following result.

Theorem 14.2.1. *One has*

$$(I-V)^{-1} = \int_{[0,T]}^{\leftarrow} (I + V\,dQ(s)).$$

14.3 Differential delay equations

Consider the equation

$$\dot{y}(t) = \int_0^\eta A(t,s)y(t-s)ds + \sum_{k=0}^m B_k(t)y(t-h_k) + f(t) \ \ (t \geq 0;\ m < \infty) \quad (3.1)$$

where $f \in C(0,T)$, and $0 \leq h_0 < h_1 < \cdots < h_m \leq \eta < \infty$ are constants, $B_k(t)$ are continuous matrices and $A(t,s)$ is Riemann integrable in s and continuous in t. Take the zero initial condition

$$y(t) = 0,\ t \leq 0 \quad (3.2)$$

and integrate (3.1). Then

$$y(t) = \int_0^t \left[\int_0^\eta A(t_1,s)y(t_1-s)ds + \sum_{k=0}^m B_k(t_1)y(t_1-h_k)\right] dt_1 + f_1(t_1), \quad (3.3)$$

where

$$f_1(t) = \int_0^t f(t_1)dt_1.$$

Take into account that according to (3.2),

$$\int_0^\eta A(t,s)y(t-s)ds = \int_0^t A(t,s)y(t-s)ds, t \le \eta\,.$$

Put

$$A_1(t,s) := \begin{cases} A(t,s) & \text{if} \quad s \le \eta\,, \\ 0 & \text{if} \quad s > \eta\,. \end{cases}$$

Then

$$\int_0^\eta A(t,s)y(t-s)ds = \int_0^t A_1(t,s)y(t-s)ds.$$

Hence,

$$\int_0^t \int_0^\eta A(t_1,s)y(t_1-s)dsdt_1 = \int_0^t \int_0^{t_1} A_1(t_1,t_1-\tau)y(\tau)d\tau dt_1$$
$$= \int_0^t K_1(t,\tau)y(\tau)d\tau,$$

where

$$K_1(t,\tau) = \int_\tau^t A_1(t_1,t_1-\tau)dt_1.$$

Moreover,

$$\int_0^t B_k(t_1)y(t_1-h_k)dt_1 = \int_{h_k}^t B_k(t_1)y(t_1-h_k)dt_1$$
$$= \int_0^{t-h_k} B_k(\tau+h_k)y(\tau)d\tau = \int_0^t \hat{B}_k(t,\tau)y(\tau)d\tau,$$

where

$$\hat{B}_k(t,\tau) := \begin{cases} B_k(\tau+h_k) & \text{if} \quad \tau \le t-h_k, \\ 0 & \text{if} \quad \tau > t-h_k. \end{cases}$$

Thus equation (3.1) can be written as equation (2.1) with $f = f_1$ and

$$K(t,\tau) = K_1(t,\tau) + \sum_{k=0}^m \hat{B}_k(t,\tau).$$

Now we can directly apply Theorem 14.2.1.

14.4 Comments

The results of this chapter are probably new.

Appendix A
The General Form of Causal Operators

The aim of this appendix is to establish the general form of a linear bounded causal operator acting in space $C(0,T)$ of continuous real functions defined on a finite segment $[0,T]$ with the sup-norm $\|.\|_{C(0,T)}$.

Let Σ_T be the σ-algebra of the Borel subsets of $[0,T]$.

Lemma A.1. *Let A be a bounded linear operator acting in $C(0,T)$. Then there is a scalar function m defined on $[0,T]\times\Sigma_T$ additive and having bounded variation* $\operatorname{var} m(t,.)$ *with respect to the second argument, such that*

$$m(.,\Delta)\in C(0,T)\ (\Delta\in\Sigma_T)\quad \textit{and}\quad \sup_{t\in[0,T]}\operatorname{var} m(t,.)\le\|A\|_{C(0,T)} \tag{1}$$

and

$$[Af](t)=\int_0^T f(s)m(t,ds)\ (0\le t\le T;\ f\in C(0,T)). \tag{2}$$

Proof. Let

$$0=t_0<t_1<\cdots<t_n=T,\ \Delta_k=t_k-t_{k-1}\ (k=1,\dots,n).$$

Take a piece-wise constant function,

$$f_n=\sum_{k=1}^n c_k\chi(\Delta_k),$$

where χ is the characteristic function:

$$(\chi(\Delta))(t)=\begin{cases}1 & \text{if } t\in(\Delta),\\ 0 & \text{if } t\notin(\Delta)\end{cases}$$

for any $\Delta\in\Sigma_T$. It is clear that A can be defined on piece-wise constant functions.

Then

$$(Af_n)(t)=\sum_{k=1}^n c_k(A\chi(\Delta_k))(t)=\sum_{k=1}^n c_k m(t,\Delta_k),$$

where $m(t,\Delta) = (A\chi(\Delta))(t)$. Hence letting $\max_k |t_k - t_{k-1}| \to 0$, we get (2). Clearly,

$$\sum_{k=1}^{n} m(t,\Delta_k) \le \|A\|_{C(0,T)} \| \sum_{k=1}^{n} \chi(\Delta_k)\|_{C(0,T)} = \|A\|_{C(0,T)} \quad (t \in (0,T)).$$

So variation of $m(t,.)$ is less than or is equal to $\|A\|_{C(0,T)}$. This completes the proof. □

Note that a result similar to the preceding lemma is well known [16].

Put $\nu(t,s) = m(t,[0,s])$. So $\nu(t,s) = m(t,[0,s]) = (A\chi[0,s])(t)$. Now (2) can be written as

$$[Af](t) = \int_0^T f(s)d_s\nu(t,s). \tag{3}$$

If, in addition, A is positive then $\nu(t,s)$ is non-decreasing in s.

Let us turn now to the causal mappings. For all $\tau \in (0,T)$ and $w \in C(0,T)$, let the projections P_τ be defined by

$$(P_\tau w)(t) = \begin{cases} w(t) & \text{if } 0 \le t \le \tau, \\ 0 & \text{if } \tau < t \le T. \end{cases}$$

In addition, $P_T = I$, $P_0 = 0$.

We have $P_\tau C(0,T) = C(0,\tau)$. Clearly, for any $w \in C(0,T)$, the function $P_\tau w$ is in $B(0,T)$, where $B(0,T)$ is the Banach space of all bounded functions defined on $[0,T]$ with the same sup-norm $\|.\|_{C(0,T)}$. Since $C(0,T)$ is embedded into $B(0,T)$, the equality

$$P_\tau A P_\tau w = P_\tau A w \quad (w \in C(0,T);\ \tau \in [0,T]) \tag{4}$$

has a sense.

Recall that a bounded linear operator A is said to be causal, if (4) is fulfilled (see Section 1.8).

Theorem A.2. *Let A be a bounded causal linear operator acting in $C(0,T)$. Then there is a function $\mu(t,s)$ defined on $[0,T]^2$, having bounded variation with respect to the second argument, and continuous with respect to the first argument, such that*

$$[Af](t) = \int_0^t f(s)d_s\mu(t,s) \quad (f \in C(0,T), 0 \le t \le T).$$

Proof. According to (3) we obtain,

$$[P_\tau Af](t) = \int_0^T f(s)d_s\nu_\tau(t,s),$$

and

$$[P_\tau A P_\tau f](t) = \int_0^T f(s) d_s \nu_\tau(t,s) \quad (0 \le \tau < T),$$

where $\nu_\tau(t,s) = 0$ for $t > \tau$ and $\nu_\tau(t,s) = \nu(t,s)$ for $t \le \tau$. Due to (4) we get

$$\int_0^T f(s) d_s \nu_\tau(t,s) = \int_0^\tau f(s) d_s \nu_\tau(t,s).$$

Take $t = \tau$. Then

$$\int_0^T f(s) d_s \nu_\tau(\tau,s) = \int_0^\tau f(s) d_s \nu_\tau(\tau,s).$$

Hence $\nu_\tau(\tau,s) = 0$ for $s > \tau$. So taking $\mu(t,s) = \nu_t(\tau,s)$ we prove the theorem. □

Theorem A.2 appears in [40].

Appendix B
Infinite Block Matrices

This appendix contains the proofs of the results applied in Chapter 7. It deals with infinite block matrices having compact off diagonal parts. Bounds for the spectrum are established and estimates for the norm of the resolvent are proposed. The main tool in this appendix is the so-called π-triangular operators defined below. The appendix is organized as follows. It consists of 7 sections. In this section we define the π-triangular operators. In Section B.2 some properties of Volterra operators are considered. In Section B.3 we establish the norm estimates and multiplicative representation for the resolvents of π-triangular operators. Section B.4 is devoted to perturbations of block triangular matrices. Bounds for the spectrum of an infinite block matrix close to a triangular infinite block matrix are presented in Section B.5. Diagonally dominant infinite block matrices are considered in Section B.6. Section B.7 contains examples illustrating our results.

B.1 Definitions

Let H be a separable complex Hilbert space, with the norm $\|.\|$ and unit operator I. All the operators considered in this appendix are linear and bounded. Recall that $\sigma(A)$ and $R_\lambda(A) = (A - \lambda I)^{-1}$ denote the spectrum and resolvent of an operator A, respectively.

Recall also that a linear operator V is called quasinilpotent, if $\sigma(V) = 0$. A linear operator is called *a Volterra operator*, if it is compact and quasinilpotent.

In what follows

$$\pi = \{P_k,\ k = 0, 1, 2, \dots\}$$

is an infinite chain of orthogonal projections P_k in H, such that

$$0 = P_0 H \subset P_1 H \subset P_2 H \subset \cdots$$

and $P_n \to I$ in the strong topology as $n \to \infty$. In addition

$$\sup_{k\geq 1} \dim \Delta\, P_k H < \infty, \quad \text{where} \quad \Delta P_k = P_k - P_{k-1}.$$

Let a linear operator A acting in H satisfy the relations

$$AP_k = P_k A P_k \ \ (k = 1, 2, \dots). \tag{1.1}$$

That is, P_k are invariant projections for A. Put

$$D := \sum_{k=1}^{\infty} \Delta P_k A \Delta P_k$$

and $V = A - D$. Then

$$A = D + V, \tag{1.2}$$

and

$$DP_k = P_k D \ \ (k = 1, 2, \dots), \tag{1.3}$$

and

$$P_{k-1} V P_k = V P_k \ (k = 2, 3, \dots); \ VP_1 = 0. \tag{1.4}$$

Definition B.1.1. Let relations (1.1)–(1.4) hold with a compact operator V. Then we will say that A is a π-triangular operator, D is a π-diagonal operator and V is a π-Volterra operator.

Besides, relation (1.2) will be called the π-triangular representation of A, and D and V will be called the π-diagonal part and π-nilpotent part of A, respectively.

B.2 Properties of π-Volterra operators

Lemma B.2.1. *Let $\pi_m = \{Q_k, k = 1, \dots, m;\ m < \infty\}$, $Q_m = I$ be a finite chain of projections. Then any operator V satisfying the condition $Q_{k-1} V Q_k = VQ_k$ $(k = 2, \dots, m)$, $VQ_1 = 0$ is a nilpotent operator. Namely, $V^m = 0$.*

Proof. Since

$$V^m = V^m Q_m = V^{m-1} Q_{m-1} V = V^{m-2} Q_{m-2} V Q_{m-1} V = \cdots$$
$$\cdots = VQ_1 \cdots VQ_{m-2} V Q_{m-1} V,$$

we have $V^m = 0$, as claimed. □

Lemma B.2.2. *Let V be a π-Volterra operator (i.e., it is compact and satisfies (1.4)). Then V is quasinilpotent.*

Proof. Thanks to the definition of a π-Volterra operator and the previous lemma, V is a limit of nilpotent operators in the operator norm. This and Theorem I.4.1 from [65] prove the lemma. □

Lemma B.2.3. *Let V be a π-Volterra operator and B be π-triangular. Then VB and BV are π-Volterra operators.*

Proof. It is obvious that

$$P_{k-1}BVP_k = P_{k-1}BP_{k-1}VP_k = BP_{k-1}VP_k = BVP_k.$$

Similarly $P_{k-1}VBP_k = VBP_k$. This proves the lemma. □

Lemma B.2.4. *Let A be a π-triangular operator. Let V and D be the π-nilpotent and π-diagonal parts of A, respectively. Then for any regular point λ of D, the operators $VR_\lambda(D)$ and $R_\lambda(D)V$ are π-Volterra ones.*

Proof. Since $P_kR_\lambda(D) = R_\lambda(D)P_k$, the previous lemma ensures the required result. □

Let Y be *a norm ideal of compact linear operators in H*. That is, Y is algebraically a two-sided ideal, which is complete in an auxiliary norm $|\cdot|_Y$ for which $|CB|_Y$ and $|BC|_Y$ are both dominated by $\|C\||B|_Y$.

In the sequel we suppose that there are positive numbers θ_k ($k \in \mathbf{N}$), with

$$\theta_k^{1/k} \to 0 \text{ as } k \to \infty,$$

such that

$$\|V^k\| \leq \theta_k |V|_Y^k \tag{2.1}$$

for an arbitrary Volterra operator

$$V \in Y. \tag{2.2}$$

Recall that the Schatten–von Neumann ideal SN_{2p} ($p = 1, 2, \dots$) is the ideal of compact operators with the finite ideal norm

$$N_{2p}(K) = [\text{Trace}(K^*K)^p]^{1/2p} \quad (K \in SN_{2p}).$$

Let $V \in SN_{2p}$ be a Volterra operator. Then due to Corollary 6.9.4 from [31], we get

$$\|V^j\| \leq \theta_j^{(p)} N_{2p}^j(V) \quad (j = 1, 2, \dots), \tag{2.3}$$

where

$$\theta_j^{(p)} = \frac{1}{\sqrt{[j/p]!}}$$

and $[x]$ means the integer part of a positive number x. Inequality (2.3) can be written as

$$\|V^{kp+m}\| \leq \frac{N_{2p}^{pk+m}(V)}{\sqrt{k!}} \quad (k = 0, 1, 2, \dots;\ m = 0, \dots, p-1). \tag{2.4}$$

In particular, if V is a Hilbert–Schmidt operator, then

$$\|V^j\| \leq \frac{N_2^j(V)}{\sqrt{j!}} \quad (j = 0, 1, 2, \dots). \tag{2.5}$$

B.3 Resolvents of π-triangular operators

Lemma B.3.1. *Let A be a π-triangular operator. Then $\sigma(A) = \sigma(D)$, where D is the π-diagonal part of A. Moreover,*

$$R_\lambda(A) = R_\lambda(D) \sum_{k=0}^{\infty} (VR_\lambda(D))^k (-1)^k. \tag{3.1}$$

Proof. Let λ be a regular point of operator D. According to the triangular representation (1.2) we obtain

$$R_\lambda(A) = (D + V - \lambda I)^{-1} = R_\lambda(D)(I + VR_\lambda(D))^{-1}.$$

Operator $VR_\lambda(D)$ for a regular point λ of operator D is a Volterra one due to Lemma B.2.3. Therefore,

$$(I + VR_\lambda(D))^{-1} = \sum_{k=0}^{\infty} (VR_\lambda(D))^k (-1)^k$$

and the series converges in the operator norm. Hence, it follows that λ is a regular point of A.

Conversely let $\lambda \notin \sigma(A)$. According to the triangular representation (1.2) we obtain

$$R_\lambda(D) = (A - V - \lambda I)^{-1} = R_\lambda(A)(I - VR_\lambda(A))^{-1}.$$

Since V is a p-Volterra, for a regular point λ of A, operator $VR_\lambda(A)$ is a Volterra one due to Lemma B.2.4. So

$$(I - VR_\lambda(A))^{-1} = \sum_{k=0}^{\infty} (VR_\lambda(A))^k$$

and the series converges in the operator norm. Thus,

$$R_\lambda(D) = R_\lambda(A) \sum_{k=0}^{\infty} (VR_\lambda(A))^k.$$

Hence, it follows that λ is a regular point of D. This finishes the proof. □

With $V \in Y$, introduce the function

$$\zeta_Y(x, V) := \sum_{k=0}^{\infty} \theta_k |V|_Y^k x^{k+1} \quad (x \geq 0).$$

Corollary B.3.2. *Let A be a π-triangular operator and let its π-nilpotent part V belong to a norm ideal Y with the property* (2.1). *Then*

$$\|R_\lambda(A)\| \leq \zeta_Y(\|R_\lambda(D)\|, V) := \sum_{k=0}^{\infty} \theta_k |V|_Y^k \|R_\lambda(D)\|^{k+1}$$

for all regular λ of A.

Indeed, according to Lemma B.2.3 and (2.1),

$$\|(VR_\lambda(D))^k\| \leq \theta_k |VR_\lambda(D)|_Y^k.$$

But

$$|VR_\lambda(D)|_Y \leq |V|_Y \|R_\lambda(D)\|.$$

Now the required result is due to (3.1).

Corollary B.3.2 and inequality (2.3) yield

Corollary B.3.3. *Let A be a π-triangular operator and its π-nilpotent part $V \in SN_{2p}$ for some integer $p \geq 1$. Then*

$$\|R_\lambda(A)\| \leq \sum_{k=0}^{\infty} \theta_k^{(p)} N_{2p}^k(V) \|R_\lambda(D)\|^{k+1} \ (\lambda \notin \sigma(A)),$$

where D is the π-diagonal part of A. In particular, if V is a Hilbert–Schmidt operator, then

$$\|R_\lambda(A)\| \leq \sum_{k=0}^{\infty} \frac{N_2^k(V)}{\sqrt{k!}} \|R_\lambda(D)\|^{k+1} \quad (\lambda \notin \sigma(A)).$$

Note that under the condition $V \in SN_{2p}$, $p > 1$, inequality (2.4) implies

$$\|R_\lambda(A)\| \leq \sum_{j=0}^{p-1} \sum_{k=0}^{\infty} \frac{N_{2p}^{pk+j}(V)}{\sqrt{k!}} \|R_\lambda(D)\|^{pk+j+1}. \tag{3.2}$$

Thanks to the Schwarz inequality, for all $x > 0$ and $a \in (0,1)$,

$$\left[\sum_{k=0}^{\infty} \frac{x^k}{\sqrt{k!}}\right]^2 = \left[\sum_{k=0}^{\infty} \frac{x^k a^k}{a^k \sqrt{k!}}\right]^2 \leq \sum_{k=0}^{\infty} a^{2k} \sum_{k=0}^{\infty} \frac{x^{2k}}{a^{2k} k!} = (1-a^2)^{-1} e^{x^2/a^2}.$$

In particular, take $a^2 = 1/2$. Then

$$\sum_{k=0}^{\infty} \frac{x^k}{\sqrt{k!}} \leq \sqrt{2} e^{x^2}.$$

Now (3.2) implies

Corollary B.3.4. *Let A be a π-triangular operator and its π-nilpotent part $V \in SN_{2p}$ for some integer $p \geq 1$. Then*

$$\|R_\lambda(A)\| \leq \zeta_p(\|R_\lambda(D)\|, V) \quad (\lambda \notin \sigma(A)),$$

where

$$\zeta_p(x, V) := \sqrt{2} \sum_{j=0}^{p-1} N_{2p}^j(V) x^{j+1} \ \exp[\, N_{2p}^{2p}(V) x^{2p}] \ (x > 0). \tag{3.3}$$

Lemma B.3.5. *Let A be a π triangular operator, whose π-nilpotent part V belongs to a norm ideal Y with the property* (2.1). *Let B be a bounded operator in H. Then for any $\mu \in \sigma(B)$ either $\mu \in \sigma(D)$, or*

$$\|A - B\|\zeta_Y(\|R_\mu(D)\|, V) \geq 1.$$

In particular, if $V \in SN_{2p}$ for some integer $p \geq 1$, then this inequality holds with $\zeta_Y = \zeta_p$.

Indeed this result follows from Corollaries B.3.2 and B.3.4.

Let us establish the multiplicative representation for the resolvent of a π-triangular operator. To this end, for bounded linear operators $X_1, X_2, \ldots, X_m$ and $j < m$, denote

$$\prod_{j \leq k \leq m}^{\rightarrow} X_k := X_j X_{j+1} \ldots X_m.$$

In addition

$$\prod_{j \leq k \leq \infty}^{\rightarrow} X_k := \lim_{m \to \infty} \prod_{j \leq k \leq m}^{\rightarrow} X_k$$

if the limit exists in the operator norm.

Lemma B.3.6. *Let $\pi = \{P_k\}_{k=1}^{\infty}$ be a chain of orthogonal projections, V a π-Volterra operator. Then*

$$(I - V)^{-1} = \prod_{k=2,3,\ldots}^{\rightarrow} (I + V\Delta P_k). \tag{3.4}$$

Proof. First let $\pi = \{P_1, \ldots, P_m\}$ be finite. According to Lemma B.2.1,

$$(I - V)^{-1} = \sum_{k=0}^{m-1} V^k. \tag{3.5}$$

On the other hand,

$$\prod_{2 \leq k \leq m}^{\rightarrow} (I + V\Delta P_k) = I + \sum_{k=2}^{m} V_k + \sum_{2 \leq k_1 < k_2 \leq m} V_{k_1} V_{k_2} + \cdots + V_2 V_3 \ldots V_m.$$

Here, as above, $V_k = V\Delta P_k$. However,

$$\begin{aligned}\sum_{2 \leq k_1 < k_2 \leq m} V_{k_1} V_{k_2} &= V \sum_{2 \leq k_1 < k_2 \leq m} \Delta P_{k_1} V \Delta P_{k_2} \\ &= V \sum_{3 \leq k_2 \leq m} P_{k_2 - 1} V \Delta P_{k_2} = V^2 \sum_{3 \leq k_2 \leq m} \Delta P_{k_2} = V^2.\end{aligned}$$

Similarly,

$$\sum_{2\leq k_1<k_3\cdots<k_j\leq m} V_{k_1}V_{k_2}\dots V_{k_j} = V^j$$

for $j < m$. Thus from (3.5) the relation (3.4) follows. The rest of the proof is left to the reader. □

Theorem B.3.7. *For any π-triangular operator A and a regular $\lambda \in \mathbb{C}$ we have*

$$R_\lambda(A) = (D-\lambda I)^{-1} \prod_{2\leq k\leq\infty}^{\rightarrow} (I - V\Delta P_k(D-\lambda I)^{-1}\Delta P_k),$$

where D and V are the π-diagonal and π-nilpotent parts of A, respectively.

Proof. Due to Lemma B.2.4, $VR_\lambda(D)$ is π-nilpotent. Now the previous lemma implies

$$(I + VR_\lambda(D))^{-1} = \prod_{2\leq k\leq m}^{\rightarrow} (I - VR_\lambda(D)\Delta P_k).$$

But $R_\lambda(D)\Delta P_k = \Delta P_k R_\lambda(D)$. This proves the result. □

Let A be a π-triangular operator and

$$\Pi(A,\lambda) := \|R_\lambda(D)\| \prod_{k=2}^{\infty}(1 + \|R_\lambda(D)\Delta P_k\|\|V\Delta P_k\|) < \infty.$$

Then from the previous theorem follows the inequality $\|R_\lambda(A)\| \leq \Pi(A,\lambda)$.

B.4 Perturbations of block triangular matrices

Let $H = l^2(\mathbb{C}^n)$ be the space of sequences $h = \{h_k \in \mathbb{C}^n\}_{k=1}^{\infty}$ with values in the Euclidean space $\mathbb{C}^n$ and the norm

$$|h|_{l^2(\mathbb{C}^n)} = \left[\sum_{k=1}^{\infty} \|h_k\|_n^2\right]^{1/2},$$

where $\|\cdot\|_n$ is the Euclidean norm in $\mathbb{C}^n$.

Consider the operator defined in $l^2(\mathbb{C}^n)$ by the upper block triangular matrix

$$A_+ = \begin{pmatrix} A_{11} & A_{12} & A_{13} & \dots \\ 0 & A_{22} & A_{23} & \dots \\ 0 & 0 & A_{33} & \dots \\ . & . & . & \dots \end{pmatrix}, \tag{4.1}$$

where A_{jk} are $n \times n$-matrices.

So $A_+ = \tilde{D} + V_+$, where V_+ and $\tilde{D}$ are the strictly upper triangular, and diagonal parts of A_+, respectively:

$$V_+ = \begin{pmatrix} 0 & A_{12} & A_{13} & A_{14} & \dots \\ 0 & 0 & A_{23} & A_{24} & \dots \\ 0 & 0 & 0 & A_{34} & \dots \\ . & . & . & . & \dots \end{pmatrix}$$

and $\tilde{D} = \operatorname{diag}[A_{11}, A_{22}, A_{33}, \dots]$. Put

$$\eta_n(\lambda) := \sup_k \|R_\lambda(A_{kk})\|_n.$$

Lemma B.4.1. *Let A_+ be the block triangular matrix defined by (4.1) and V_+ be a compact operator belonging to a norm ideal Y with the property (2.1). Then $\sigma(A_+) = \sigma(\tilde{D})$, and*

$$|R_\lambda(A_+)|_{l^2} \le \zeta_Y(\eta_n(\lambda), V_+) \tag{4.2}$$

for all regular λ of $\tilde{D}$. Moreover, for any bounded operator B acting in $l^2(\mathbb{C}^n)$ and a $\mu \in \sigma(B)$, either $\mu \in \sigma(\tilde{D})$, or

$$q\zeta_Y(\eta_n(\mu), V_+) \ge 1$$

where $q := |A_+ - B|_{l^2(\mathbb{C}^n)}$. In particular, if

$$V_+ \in SN_{2p} \ (p = 1, 2, \dots) \tag{4.3}$$

then $\zeta_Y = \zeta_p$.

Proof. Let P_j, $j = 0, 1, 2, \dots$ be projections onto the subspaces of $l^2(\mathbb{C}^n)$ generated by the first nj elements of the standard basis. Then $\pi = \{P_k\}$ is the infinite chain of orthogonal projections in $l^2(\mathbb{C}^n)$, such that (1.1) holds and $P_n \to I$ strongly as $n \to \infty$. Moreover, $\dim \Delta P_k H \equiv n$ $(k = 1, 2, \dots)$ and

$$A_{jk} = \Delta P_j \tilde{A} \Delta P_k, \tilde{D} = \sum_{j=1}^{\infty} A_{jj}.$$

Hence it follows that A_+ is a π triangular operator. Now Corollaries B.3.2 and B.3.5 prove the result. □

Let

$$\|R_\lambda(A_{kk})\| \le \phi(\rho(A_{kk}, \lambda)) := \sum_{l=0}^{n-1} \frac{c_l}{\rho^{l+1}(A_{kk}, \lambda)} \ (\lambda \notin \sigma(\tilde{D}))$$

where c_l are non-negative coefficients, independent of k, and $\rho(A, \lambda)$ is the distance between a complex point λ and $\sigma(A)$. Then

$$\|R_\lambda(A_{kk})\| \le \phi(\rho(\tilde{D}, \lambda)) := \sum_{l=0}^{n-1} \frac{c_l}{\rho^{l+1}(\tilde{D}, \lambda)} \ (\lambda \notin \sigma(\tilde{D})) \tag{4.4}$$

and

$$\rho(\tilde{D},\lambda) = \inf_{k=1,2,\dots} \min_{j=1,\dots,n} |\lambda - \lambda_j(A_{kk})|$$

is the distance between a point λ and $\sigma(\tilde{D})$, and

$$\phi(y) = \sum_{k=0}^{n-1} \frac{c_k}{y^{k+1}} \quad (y > 0).$$

We have

$$\eta_n(\lambda) = \sup_{j=1,2,\dots} \|R_\lambda(A_{jj})\| \leq \phi(\rho(\tilde{D},\lambda)).$$

Under (4.3), Lemma B.4.1 gives us the inequality

$$|R_\lambda(A_+)|_{l^2(\mathbb{C}^n)} \leq \zeta_p(\phi(\rho(\tilde{D},\lambda)), V_+) \tag{4.5}$$

for all regular λ of $\tilde{D}$, provided $V_+ \in SN_{2p}$. So for a bounded operator B and $\mu \in \sigma((B)$, either $\mu \in \sigma(\tilde{D})$ or

$$q\zeta_p(\phi(\rho(\tilde{D},\mu)), V_+) \geq 1. \tag{4.6}$$

Furthermore, let $C = (c_{jk})_{j,k=1}^n$ be an $n \times n$-matrix. Then as it is shown in Section 2.3,

$$\|R_\lambda(C)\|_n \leq \sum_{k=0}^{n-1} \frac{g^k(C)}{\sqrt{k!}\rho^{k+1}(C,\lambda)},$$

where $\lambda_k(C)$; $k = 1,\dots,n$ are the eigenvalues of C including their multiplicities, and

$$g(C) = (N_2^2(C) - \sum_{k=1}^{n} |\lambda_k(C)|^2)^{1/2}.$$

In particular, the inequalities

$$g^2(C) \leq N_2^2(C) - |\operatorname{Trace} C^2| \quad \text{and} \quad g^2(C) \leq \frac{1}{2} N_2^2(C^* - C) \tag{4.7}$$

are true. If C is a normal matrix, then $g(C) = 0$. Thus

$$\|R_\lambda(A_{jj})\|_n \leq \sum_{k=0}^{n-1} \frac{g^k(A_{jj})}{\sqrt{k!}\rho^{k+1}(A_{jj},\lambda)}.$$

Since $\tilde{D}$ is bounded, we have

$$g_0 := \sup_{k=1,2,\dots} g(A_{kk}) < \infty.$$

Then one can take $\phi(y) = \phi_0(y)$, where

$$\phi_0(y) = \sum_{k=0}^{n-1} \frac{g_0^k}{\sqrt{k!}y^{k+1}}.$$

If all the diagonal matrices A_{kk} are normal, then $g(A_{kk}) = 0$ and $\phi_0(y) = 1/y$. Relation (4.6) yields

Lemma B.4.2. *Let A_+ be defined by* (4.1) *and B a linear operator on $l^2(\mathbb{C}^n)$. If, in addition, condition* (4.3) *holds, then for any $\mu \in \sigma(B)$, there is a $\lambda \in \sigma(\tilde{D})$, such that*

$$|\lambda - \mu| \le r_p(q, V_+)),$$

where $r_p(q, V_+)$ is the unique positive root of the equation

$$q\zeta_p(\phi_0(y), V_+) = 1. \tag{4.8}$$

Here y is the unknown.

It is simple to see that

$$r_p(q, V_+) \le y_n(z_p(q, V_+)), \tag{4.9}$$

where $y_n(b)$ is the unique positive root of the equation

$$\phi_0(y) = b \ \ (b = \text{const} > 0),$$

and $z_p(q, V_+)$ is the unique positive root of the equation

$$q\zeta_p(x, V_+) = q\sqrt{2}\sum_{j=0}^{p-1} N_{2p}^j(V_+)x^{j+1} \ \exp[\ N_{2p}^{2p}(V_+)x^{2p}] = 1. \tag{4.10}$$

Furthermore, due to Lemma 2.4.3, $y_n(b) \le p_n(b)$, where

$$p_n(b) = \begin{cases} \phi_0(1)/b & \text{if } \phi_0(1) \ge b, \\ \sqrt[n]{\phi_0(1)/b} & \text{if } \phi_0(1) < b. \end{cases}$$

We need the following

Lemma B.4.3. *The unique positive root z_a of the equation*

$$\sum_{j=0}^{p-1} z^{j+1} \ \exp[z^{2p}] = a \ \ (a = \text{const} > 0) \tag{4.11}$$

satisfies the inequality $z_a \ge \delta_p(a)$, where

$$\delta_p(a) := \begin{cases} a/pe & \text{if } \ a \le pe, \\ [\ln\ (a/p)]^{1/2p} & \text{if } \ a > pe. \end{cases} \tag{4.12}$$

For the proof see [31, Lemma 8.3.2]. Put in (4.10) $N_{2p}(V_+)x = z$. Then we get equation (4.12) with $a = N_{2p}(V_+)/q\sqrt{2}$. The previous lemma implies

$$z_p(q, V_+) \geq \gamma_p(q, V_+), \quad \text{where} \quad \gamma_p(q, V_+) := \frac{\delta_p(N_{2p}(V_+)/q\sqrt{2})}{N_{2p}(V_+)}.$$

We thus get

$$r_p(q, V_+) \leq p_n(\gamma_p(q, V_+)). \tag{4.13}$$

B.5 Block matrices close to triangular ones

Consider in $l^2(\mathbb{C}^n)$ the operator defined by the block matrix

$$\tilde{A} = \begin{pmatrix} A_{11} & A_{12} & A_{13} & \dots \\ A_{21} & A_{22} & A_{23} & \dots \\ A_{31} & A_{32} & A_{33} & \dots \\ . & . & . & \dots \end{pmatrix} \tag{5.1}$$

where A_{jk} are $n \times n$-matrices. Clearly,

$$\tilde{A} = \tilde{D} + V_+ + V_-$$

where V_+ is the strictly upper triangular, part, $\tilde{D}$ is the diagonal part and V_- is the strictly lower triangular part of $\tilde{A}$:

$$V_- = \begin{pmatrix} 0 & 0 & 0 & 0 & \dots \\ A_{21} & 0 & 0 & 0 & \dots \\ A_{31} & A_{32} & 0 & 0 & \dots \\ A_{41} & A_{42} & A_{43} & 0 & \dots \\ . & . & . & \dots & . \end{pmatrix}.$$

Now we get the main result of the paper which is due to (4.6) with $B = \tilde{A}$.

Recall that ϕ_0 is defined in the previous section and ζ_p is defined by (3.3).

Theorem B.5.1. *Let $\tilde{A}$ be defined by* (5.1) *and condition* (4.3) *hold. Then for any $\mu \in \sigma(\tilde{A})$, either $\mu \in \sigma(\tilde{D})$, or there is a $\lambda \in \sigma(\tilde{D})$, such that*

$$|V_-|_{l^2(\mathbb{C}^n)} \zeta_p(\phi_0(|\lambda - \mu|), V_+) \geq 1.$$

The theorem is exact in the following sense: if $V_- = 0$, then $\sigma(\tilde{A}) = \sigma(\tilde{D})$. Moreover, Lemma B.4.2 with $B = \tilde{A}$ implies

Corollary B.5.2. *Let $\tilde{A}$ be defined by* (5.1) *and condition* (4.3) *hold. Then for any $\mu \in \sigma(\tilde{D})$, there is a $\lambda \in \sigma(\tilde{D})$, such that*

$$|\lambda - \mu| \leq r_p(\tilde{A}),$$

where $r_p(\tilde{A})$ *is the unique positive root of the equation*

$$|V_-|_{l^2(\mathbb{C}^n)}\zeta_p(\phi_0(y), V_+) = 1.$$

Moreover, (4.13) *gives us the bound for* $r_p(\tilde{A})$ *if we take* $q = |V_-|_{l^2(\mathbb{C}^n)}$.

Note that in Theorem B.5.1 it is enough that V_+ is compact. Operator V_- can be noncompact.

Clearly, one can exchange V_+ and V_-.

B.6 Diagonally dominant block matrices

Put

$$m_{jk} = \|A_{jk}\|_n \quad (j, k = 1, 2, \dots)$$

and consider the matrix

$$M = (m_{jk})_{j,k=1}^{\infty}.$$

Lemma B.6.1. *The spectral radius* $r_s(\tilde{A})$ *of* $\tilde{A}$ *is less than or is equal to the spectral radius of* M.

Proof. Let $A_{jk}^{(\nu)}$ and $m_{jk}^{(\nu)}$ $(\nu = 2, 3, \dots)$ be the entries of $\tilde{A}^\nu$ and M^ν, respectively. We have

$$\|A_{jk}^{(2)}\|_n = \|\sum_{l=1}^{\infty} A_{jl}A_{lk}\|_n \le \sum_{l=1}^{\infty} \|A_{jl}\|_n\|A_{lk}\|_n = \sum_{l=1}^{\infty} m_{jl}m_{lk} = m_{jk}^{(2)}.$$

Similarly, we get $\|A_{jk}^{(\nu)}\|_n \le m_{jk}^{(\nu)}$.

But for any $h = \{h_k\} \in l^2(\mathbb{C}^n)$, we have

$$|\tilde{A}h|^2_{l^2(C^n)} \le \sum_{j=1}^{\infty}\left(\sum_{k=1}^{\infty}\|A_{jk}h_k\|_n\right)^2 \le \sum_{j=1}^{\infty}\left(\sum_{k=1}^{\infty} m_{jk}\|h_k\|_n\right)^2 = |M\tilde{h}|^2_{l^2(\mathbb{R})}$$

where

$$\tilde{h} = \{\|h_k\|_n\} \in l^2(R^1).$$

Since

$$|h|^2_{l^2(C^n)} = \sum_{k=1}^{\infty}\|h_k\|_n^2 = |\tilde{h}|^2_{l^2(\mathbb{R})},$$

we obtain $|\tilde{A}^\nu|_{l^2(C^n)} \le |M^\nu|_{l^2(\mathbb{R})}$ $(\nu = 2, 3, \dots)$. Now the Gel'fand formula for the spectral radius yields the required result. □

Let

$$S_j := \sum_{k=1, k\ne j}^{\infty} \|A_{jk}\|_n.$$

Theorem B.6.2. *Let A_{jj} be invertible for all integer j. In addition, let*

$$\sup_j S_j < \infty \tag{6.1}$$

and there be an $\epsilon > 0$, such that

$$\|A_{jj}^{-1}\|_n^{-1} - S_j \geq \epsilon \ (j = 1, 2, \dots). \tag{6.2}$$

Then $\tilde{A}$ is invertible. Moreover, let

$$\psi(\lambda) := \sup_j \|(A_{jj} - \lambda)^{-1}\| S_j < 1.$$

Then λ is a regular point of $\tilde{A}$, and

$$|(\tilde{A} - \lambda)^{-1}|_{l^2(\mathbb{C}^n)} \leq (1 - \psi(\lambda))^{-1} \sup_j \|(A_{jj} - \lambda)^{-1}\|_n.$$

Proof. Put $W = \tilde{A} - \tilde{D} = V_+ + V_-$ with an invertible $\tilde{D}$. That is, W is the off-diagonal part of $\tilde{A}$, and

$$\tilde{A} = \tilde{D} + W = \tilde{D}(I + \tilde{D}^{-1}W). \tag{6.3}$$

Clearly,

$$\sum_{k=1, k\neq j}^{\infty} \|A_{jj}^{-1}A_{jk}\|_n \leq S_j\|A_{jj}^{-1}\|.$$

From (6.2) it follows that

$$1 - S_j\|A_{jj}^{-1}\|_n \geq \|A_{jj}^{-1}\|_n\epsilon \ (j = 1, 2, \dots).$$

Therefore

$$\sup_j \sum_{k=1, k\neq j}^{\infty} \|A_{jj}^{-1}A_{jk}\|_n < 1.$$

Then thanks to Lemma 2.4.8 on the bound for the spectral radius and the previous lemma the spectral radius $r_s(\tilde{D}^{-1}W)$ of the matrix $\tilde{D}^{-1}W$ is less than one. Therefore $I + \tilde{D}^{-1}W$ is invertible. Now (6.3) implies that $\tilde{A}$ is invertble. This proves the theorem. $\square$

It should be noted that condition (6.1) implies that the off-diagonal part W of $\tilde{A}$ is compact, since under (6.1) the sequence of the finite-dimensional operators

$$W_l := \begin{pmatrix} 0 & A_{12} & A_{13} & \dots & A_{1l} \\ A_{21} & 0 & A_{23} & \dots & A_{2l} \\ A_{31} & A_{32} & 0 & \dots & A_{3l} \\ . & . & . & \dots & . \\ A_{l1} & A_{l2} & A_{l3} & \dots & 0 \end{pmatrix}$$

converges to W in the norm of space $l^2(\mathbb{C}^n)$ as $l \to \infty$.

B.7 Examples

Let $n = 2$. Then $\tilde{D}$ is the orthogonal sum of the 2×2-matrices

$$A_{kk} = \begin{pmatrix} a_{2k-1,2k-1} & a_{2k-1,2k} \\ a_{2k,2k-1} & a_{2k,2k} \end{pmatrix} \quad (k = 1, 2, \dots).$$

If A_{kk} are real matrices, then due to the above-mentioned inequality $g^2(C) \leq N_2^2(C^* - C)/2$, we have

$$g(A_{kk}) \leq |a_{2k-1,2k} - a_{2k,2k-1}|.$$

So one can take

$$\phi_0(y) = \frac{1}{y}\left(1 + \frac{\tilde{g}_0}{y}\right) \quad \text{with} \quad \tilde{g}_0 := \sup_k |a_{2k-1,2k} - a_{2k,2k-1}|.$$

Besides, $\sigma(\tilde{D}) = \{\lambda_{1,2}(A_{kk})\}_{k=1}^{\infty}$, where

$$\lambda_{1,2}(A_{kk}) = \frac{1}{2}(a_{2k-1,2k-1} + a_{2k,2k} \pm [(a_{2k-1,2k-1} - a_{2k,2k})^2 - a_{2k-1,2k}a_{2k,2k-1}]^{1/2}).$$

Now we can directly apply Theorems B.5.1 and Corollary B.5.2.

Furthermore, let $L^2(\omega, \mathbb{C}^n)$ be the space of vector-valued functions defined on a bounded subset ω of $\mathbb{R}^m$ with the scalar product

$$(f, g) = \int_\omega (f(s), g(s))_{C^n} ds$$

where $(.,.)_{C^n}$ is the scalar product in $\mathbb{C}^n$. Let us consider in $L^2(\omega, \mathbb{C}^n)$ the matrix integral operator

$$(Tf)(x) = \int_\omega K(x, s) f(s) ds$$

with the condition

$$\int_\omega \int_\omega \|K(x, s)\|_n^2 dx\, ds < \infty.$$

That is, T is a Hilbert–Schmidt operator.

Let $\{e_k(x)\}$ be an orthogonal normal basis in $L^2(\omega, \mathbb{C}^n)$ and

$$K(x, s) = \sum_{j,k=1}^{\infty} A_{jk} e_k(s) e_k(x)$$

be the Fourier expansion of K, with the matrix coefficients A_{jk}. Then T is unitarily equivalent to the operator $\tilde{A}$ defined by (1.1). Now one can apply Theorems B.5.1 and B.6.1, and Corollary B.5.2.

This appendix is based on the paper [37].

Bibliography

[1] Agarwal, R. Berezansky, L. Braverman, E. and A. Domoshnitsky, *Nonoscillation Theory of Functional Differential Equations and Applications*, Elsevier, Amsterdam, 2012.

[2] Ahiezer, N.I. and Glazman, I.M., *Theory of Linear Operators in a Hilbert Space*. Pitman Advanced Publishing Program, Boston, 1981.

[3] Aizerman, M.A., On a conjecture from absolute stability theory, *Ushechi Matematicheskich Nauk*, **4 (4)**, (1949) 187–188 In Russian.

[4] Azbelev, N.V. and Simonov, P.M., *Stability of Differential Equations with Aftereffects*, Stability Control Theory Methods Appl. v. 20, Taylor & Francis, London, 2003.

[5] Bazhenova, L.S., *The IO-stability of equations with operators causal with respect to a cone. Mosc. Univ. Math. Bull.* **57** (2002), no.3, 33–35; translation from Vestn. Mosk. Univ., Ser. I, (2002), no.3, 54–57.

[6] Berezansky, L. and Braverman, E., On exponential stability of linear differential equations with several delays, *J. Math. Anal. Appl.* **324** (2006) 1336–1355

[7] Berezansky, L. and Braverman, E., On stability of some linear and nonlinear delay differential equations, *J. Math. Anal. Appl.*, **314** (2006) 391–411.

[8] Berezansky, L. Idels, L. and Troib, L., Global dynamics of Nicholson-type delay systems with applications, *Nonlinear Analysis: Real World Applications*, **12** (2011) 436-445.

[9] Bhatia, R., *Matrix Analysis*, Springer, New York, 1997.

[10] Burton, T.A., *Stability of Periodic Solutions of Ordinary and Functional Differential Equations*, Academic Press, New York, 1985.

[11] Butcher, E.A., Ma, H., Bueler, E., Averina, V., and Szabo, Z., Stability of linear time-periodic delay differential equations via Chebyshev polynomials, *Int. J. Numer. Meth. Eng.*, **59**, (2004) 859–922.

[12] Bylov, B.F., Grobman, B.M., Nemyckii, V.V. and Vinograd, R.E., *The Theory of Lyapunov Exponents*, Nauka, Moscow, 1966 (In Russian).

[13] Corduneanu, C., *Functional Equations with Causal Operators*, Taylor and Francis, London, 2002.

[14] Daleckii, Yu.L. and Krein, M.G., *Stability of Solutions of Differential Equations in Banach Space*, Amer. Math. Soc., Providence, R.I., 1974.

[15] Drici, Z., McRae, F.A. and Vasundhara Devi, J., Differential equations with causal operators in a Banach space. *Nonlinear Anal., Theory Methods Appl.* **62**, (2005) no.2 (A), 301–313.

[16] Dunford, N. and Schwartz, J.T., *Linear Operators, part I,* Interscience Publishers, Inc., New York, 1966.

[17] Feintuch, A., Saeks, R., *System Theory. A Hilbert Space Approach.* Ac. Press, New York, 1982.

[18] Gantmakher, F.R., *Theory of Matrices.* Nauka, Moscow, 1967. In Russian.

[19] Gel'fand, I.M. and Shilov, G.E., *Some Questions of Theory of Differential Equations.* Nauka, Moscow, 1958. In Russian.

[20] Gel'fond, A.O., *Calculations Of Finite Differences*, Nauka, Moscow. 1967, In Russian.

[21] Gil', M.I., Estimates for norms of matrix-valued functions, *Linear and Multilinear Algebra*, **35** (1993), 65–73.

[22] Gil', M.I., On solvability of nonlinear equations in lattice normed spaces, *Acta Sci. Math.*, **62**, (1996), 201–215.

[23] Gil', M.I., The freezing method for evolution equations. *Communications in Applied Analysis*, **1**, (1997), no. 2, 245–256.

[24] Gil', M.I., *Stability of Finite and Infinite Dimensional Systems*, Kluwer Academic Publishers, Boston, 1998.

[25] Gil', M.I., Perturbations of simple eigenvectors of linear operators, *Manuscripta Mathematica*, **100**, (1999), 213–219.

[26] Gil', M.I., On bounded input-bounded output stability of nonlinear retarded systems, *Robust and Nonlinear Control*, **10**, (2000), 1337–1344.

[27] Gil', M.I., On Aizerman–Myshkis problem for systems with delay. *Automatica*, **36**, (2000), 1669–1673.

[28] Gil', M.I., Existence and stability of periodic solutions of semilinear neutral type systems, *Dynamics Discrete and Continuous Systems*, **7**, (2001), no. 4, 809–820.

[29] Gil', M.I., On the "freezing" method for nonlinear nonautonomous systems with delay, *Journal of Applied Mathematics and Stochastic Analysis* **14**, (2001), no. 3, 283–292.

[30] Gil', M.I., Boundedness of solutions of nonlinear differential delay equations with positive Green functions and the Aizerman–Myshkis problem, *Nonlinear Analysis, TMA*, **49** (2002) 1065–168.

[31] Gil', M.I., *Operator Functions and Localization of Spectra*, Lecture Notes In Mathematics vol. 1830, Springer-Verlag, Berlin, 2003.

[32] Gil', M.I., Bounds for the spectrum of analytic quasinormal operator pencils in a Hilbert space, *Contemporary Mathematics*, **5**, (2003), no 1, 101–118

[33] Gil', M.I., On bounds for spectra of operator pencils in a Hilbert space, *Acta Mathematica Sinica*, **19** (2003), no. 2, 313–326

[34] Gil', M.I., On positive solutions of nonlinear equations in a Banach lattice, *Nonlinear Functional Analysis and Appl.*, **8** (2003), 581–593.

[35] Gil', M.I., Bounds for characteristic values of entire matrix pencils, *Linear Algebra Appl.* **390**, (2004), 311–320

[36] Gil', M.I., The Aizerman–Myshkis problem for functional-differential equations with causal nonlinearities, *Functional Differential Equations*, **11**, (2005) no 1-2, 445–457

[37] Gil', M.I., Spectrum of infinite block matrices and π-triangular operators, *El. J. of Linear Algebra*, **16** (2007) 216–231

[38] Gil', M.I., Estimates for absolute values of matrix functions, *El. J. of Linear Algebra*, **16** (2007) 444–450

[39] Gil', M.I., Explicit stability conditions for a class of semi-linear retarded systems, *Int. J. of Control*, **322**, (2007) no. 2, 322–327.

[40] Gil', M.I., Positive solutions of equations with nonlinear causal mappings, *Positivity*, **11**, (2007), no. 3, 523–535.

[41] Gil', M.I., L^2-stability of vector equations with causal mappings, *Dynamic Systems and Applications*, **17** (2008), 201–220.

[42] Gil', M.I., Estimates for Green's Function of a vector differential equation with variable delays, *Int. J. Applied Math. Statist*, **13** (2008), 50–62.

[43] Gil', M.I., Estimates for entries of matrix valued functions of infinite matrices. *Math. Phys. Anal. Geom.* **11**, (2008), no. 2, 175–186.

[44] Gil', M.I., Inequalities of the Carleman type for Neumann–Schatten operators *Asian-European J. of Math.*, **1**, (2008), no. 2, 203–212.

[45] Gil', M.I., Upper and lower bounds for regularized determinants, *Journal of Inequalities in Pure and Appl. Mathematics*, **9**, (2008), no. 1, 1–6.

[46] Gil', M.I., *Localization and Perturbation of Zeros of Entire Functions*. Lecture Notes in Pure and Applied Mathematics, 258. CRC Press, Boca Raton, FL, 2009.

[47] Gil', M.I., Perturbations of functions of operators in a Banach space, *Math. Phys. Anal. Geom.* **13**, (2009) 69–82.

[48] Gil', M.I., Lower bounds and positivity conditions for Green's functions to second order differential-delay equations. *Electron. J. Qual. Theory Differ. Equ.*, (2009) no. 65, 1–11.

[49] Gil', M.I., L^2-absolute and input-to-state stabilities of equations with nonlinear causal mappings, *J. Robust and Nonlinear systems*, **19**, (2009), 151–167.

[50] Gil', M.I., Meromorphic functions of matrix arguments and applications *Applicable Analysis*, **88** (2009), no. 12, 1727–1738

[51] Gil', M.I., Perturbations of functions of diagonalizable matrices, *Electr. J. of Linear Algebra*, **20** (2010) 303–313.

[52] Gil', M.I. Stability of delay differential equations with oscillating coefficients, *Electronic Journal of Differential Equations*, 2010, (2010), no. 11, 1–5.

[53] Gil', M.I., Norm estimates for functions of matrices with simple spectrum, *Rendiconti del Circolo Matematico di Palermo*, 59, (2010) 215–226

[54] Gil', M.I., Stability of vector functional differential equations with oscillating coefficients, *J. of Advanced Research in Dynamics and Control Systems*, **3**, (2011), no. 1, 26–33.

[55] Gil', M.I., Stability of functional differential equations with oscillating coefficients and distributed delays, *Differential Equations and Applications*, **3** (2011), no. 11, 1–19.

[56] Gil', M.I., Estimates for functions of finite and infinite matrices. Perturbations of matrix functions. In: Albert R. Baswell (editor) *Advances in Mathematics Research*, **16**, Nova Science Publishers, Inc., New York, 2011, pp. 25–90

[57] Gil', M.I., Ideals of compact operators with the Orlicz norms *Annali di Matematica. Pura Appl.*, Published online, October, 2011.

[58] Gil', M.I., The L^p- version of the generalized Bohl–Perron principle for vector equations with delay, *Int. J. Dynamical Systems and Differential Equations*, **3**, (2011) no. 4, 448–458.

[59] Gil', M.I., The L^p-version of the generalized BohlPerron principle for vector equations with infinite delay, *Advances in Dynamical Systems and Applications*, **6** (2011), no. 2, 177–184.

[60] Gil', M.I., Stability of retarded systems with slowly varying coefficient *ESAIM: Control, Optimisation and Calculus of Variations*, published online Sept. 2011.

[61] Gil', M.I., Stability of vector functional differential equations: a survey, *Quaestiones Mathematicae*, **35** (2012), 83–131.

[62] Gil', M.I., Exponential stability of periodic systems with distributed delays, *Asian J. of Control*, (accepted for publication)

[63] Gil', M.I., Ailon, A. and Ahn, B.-H., On absolute stability of nonlinear systems with small delays, *Mathematical Problems in Engineering*, **4**, (1998) 423–435.

[64] Gohberg, I., Goldberg, S. and Krupnik, N., *Traces and Determinants of Linear Operators*, Birkhäuser Verlag, Basel, 2000.

[65] Gohberg, I. and Krein, M.G., *Introduction to the Theory of Linear Nonselfadjoint Operators*, Trans. Mathem. Monographs, v. 18, Amer. Math. Soc., Providence, R.I., 1969.

[66] Gopalsamy, K., *Stability and Oscillations in Delay Differential Equations of Population Dynamics.* Kluwer Academic Publishers, Dordrecht, 1992.

[67] Gu, K., Kharitonov, V. and Chen, J., *Stability of Time-delay Systems*, Birkhäuser, Boston, 2003.

[68] Guter, P., Kudryavtsev L. and Levitan, B., *Elements of the Theory of Functions*, Fizmatgiz, 1963. In Russian.

[69] Halanay, A., Stability Theory of Linear Periodic Systems with Delay, *Rev. Math. Pure Appl.* **6(4)**, (1961), 633–653.

[70] Halanay, A., *Differential Equations: Stability, Oscillation, Time Lags*, Academic Press, NY, 1966.

[71] Hale, J.K., *Theory of Functional Differential Equations*, Springer- Verlag, New York, 1977.

[72] Hale, J.K. and Lunel, S.M.V., *Introduction to Functional Differential Equations*, Springer, New York, 1993.

[73] Hewitt, E. and Stromberg, K., *Real and Abstract Analysis*, Springer Verlag, Berlin 1969.

[74] Horn R.A and Johnson, C.R., *Matrix Analysis*, Cambridge Univ. Press, Cambridge, 1985

[75] Insperger, T. and Stepan, G., Stability of the damped Mathieu equation with time-delay, *J. Dynam. Syst., Meas. Control*, **125**, (2003) no. 2, 166–171.

[76] Izobov, N.A., Linear systems of ordinary differential equations. *Itogi Nauki i Tekhniki. Mat. Analis*, **12**, (1974) 71–146, In Russian.

[77] Kolmanovskii, V. and Myshkis, A., *Applied Theory of Functional Differential Equations*, Kluwer, Dordrecht, 1999.

[78] Kolmanovskii, V.B. and Nosov, V.R., *Stability of Functional Differential Equations*, Ac Press, London, 1986.

[79] Krasnosel'skii, A.M., *Asymptotic of Nonlinearities and Operator Equations*, Birkhäuser Verlag, Basel, 1995.

[80] Krasnosel'skij, M.A., Lifshits, J. and Sobolev A., *Positive Linear Systems. The Method of Positive Operators*, Heldermann Verlag, Berlin, 1989.

[81] Krisztin, T., On stability properties for one-dimensional functional-differential equations, *Funkcial. Ekvac.* **34** (1991) 241–256.

[82] Kurbatov, V., *Functional Differential Operators and Equations*, Kluwer Academic Publishers, Dordrecht, 1999.

[83] Lakshmikantham, V., Leela, S., Drici, Z. and McRae, F.A., *Theory of Causal Differential Equations*, Atlantis Studies in Mathematics for Engineering and Science, 5. Atlantis Press, Paris, 2009.

[84] Lampe, B.P. and Rosenwasser, E.N., Stability investigation for linear periodic time-delayed systems using Fredholm theory, *Automation and Remote Control*, **72**, (2011) no. 1, 38–60.

[85] Liao, Xiao Xin, *Absolute Stability of Nonlinear Control Systems*, Kluwer, Dordrecht, 1993.

[86] Liberzon, M.R., Essays on the absolute stability theory. *Automation and Remote Control*, **67**, (2006), no. 10, 1610–1644.

[87] Lillo, J.C., Oscillatory solutions of the equation $y'(x) = m(x)y(x - n(x))$. *J. Differ. Equations* **6**, (1969) 1–35.

[88] Liu Xinzhi, Shen Xuemin and Zhang, Yi, Absolute stability of nonlinear equations with time delays and applications to neural networks. *Math. Probl. Eng.* **7**, (2001) no.5, 413–431.

[89] Liz, E., Tkachenko, V. and Trofimchuk, S., A global stability criterion for scalar functional differential equations, *SIAM J. Math. Anal.* **35** (2003) 596–622.

[90] Marcus, M. and Minc, H., *A Survey of Matrix Theory and Matrix Inequalities*. Allyn and Bacon, Boston, 1964.

[91] Meyer-Nieberg, P., *Banach Lattices*, Springer-Verlag, 1991.

[92] Michiels, W. and Niculescu, S.I., *Stability and Stabilization of Time-Delay Systems. An Eigenvalue-Based Approach*, SIAM, Philadelphia, 2007.

[93] Mitrinović, D.S., Pecaric, J.E. and Fink, A.M., *Inequalities Involving Functions and their Integrals and Derivatives*, Kluwer Academic Publishers, Dordrecht, 1991.

[94] Myshkis, A.D., On solutions of linear homogeneous differential equations of the first order of stable type with a retarded arguments, *Mat. Sb., N. Ser.* **28(70)**, (1951) 15–54.

[95] Myshkis, A.D., On some problems of theory of differential equations with deviation argument, *Uspechi Matemat. Nauk*, **32 (194)**, (1977) no 2, 173–202. In Russian.

[96] Niculescu, S.I., *Delay Effects on Stability: a Robust Control Approach*, Lecture Notes in Control and Information Sciences, vol. 269, Springer, London, 2001.

[97] Ostrowski, A.M., Note on bounds for determinants with dominant principal diagonals, *Proc. of AMS*, **3**, (1952) 26–30.

[98] Ostrowski, A.M., *Solution of Equations in Euclidean and Banach spaces*. Academic Press, New York - London, 1973.

[99] Pietsch, A., *Eigenvalues and s-Numbers*, Cambridge University Press, Cambridge, 1987.

[100] Razvan, V., *Absolute Stability of Equations with Delay*, Nauka, Moscow, 1983. In Russian.

[101] Richard, J.-P., Time-delay systems: an overview of some recent advances and open problems, *Automatica*, **39**, (2003) 1667–1694.

[102] So, J.W.H., Yu, J.S. and Chen, M.P., Asymptotic stability for scalar delay differential equations, *Funkcial. Ekvac.* **39** (1996) 1–17.

[103] Stewart, G.W. and Sun Ji-guang, *Matrix Perturbation Theory*, 1990. Academic Press, New York.

[104] Tyshkevich, V.A., *Some Questions of the Theory of Stability of Functional Differential Equations*, Naukova Dumka, Kiev, 1981. In Russian.

[105] Vath, M., *Volterra and Integral Equations of Vector Functions*, Marcel Dekker, 2000.

[106] Vidyasagar, M., *Nonlinear Systems Analysis*, second edition. Prentice-Hall. Englewood Cliffs, New Jersey, 1993.

[107] Vinograd, R., An improved estimate in the method of freezing, *Proc. Amer. Soc.* **89 (1)**, (1983) 125–129.

[108] Vulikh, B.Z., *Introduction to the Theory of Partially Ordered Spaces*, Wolters-Noordhoff Scientific Publications LTD, Groningen, 1967.

[109] Wang, X. and Liao, L., Asymptotic behavior of solutions of neutral differential equations with positive and negative coefficients, *J. Math. Anal. Appl.* **279** (2003) 326–338.

[110] Yakubovich V.A. and Starzhinskii, V.M., *Differential Equations with Periodic Coefficients*, John Wiley, New York, 1975.

[111] Yang, Bin and Chen, Mianyun, Delay-dependent criterion for absolute stability of Luré type control systems with time delay. *Control Theory Appl*, **18**, (2001) no. 6, 929–931.

[112] Yoneyama, Toshiaki, On the stability for the delay-differential equation $\dot{x}(t) = -a(t)f(x(t - r(t)))$. *J. Math. Anal. Appl.* **120**, (1986) 271–275.

[113] Yoneyama, Toshiaki, On the 3/2 stability theorem for one-dimensional delay-differential equations. *J. Math. Anal. Appl.* **125**, (1987) 161–173.

[114] Yoneyama, Toshiaki, The 3/2 stability theorem for one-dimensional delay-differential equations with unbounded delay. *J. Math. Anal. Appl.* **165**, (1992) no.1, 133–143.

[115] Zeidler, E., *Nonlinear Functional Analysis and its Applications*, Springer-Verlag, New York, 1986.

[116] Zevin, A.A. and Pinsky M.A., A new approach to the Luré problem in the theory of exponential stability and bounds for solutions with bounded nonlinearities, *IEEE Trans. Autom. Control*, **48**, (2003) no. 10, 1799–1804.

[117] Zhang, Z. and Wang, Z., Asymptotic behavior of solutions of neutral differential equations with positive and negative coefficients, *Ann. Differential Equations* **17** (3) (2001) 295–305.

Index